THE
BACKYARD

CHICKEN
KEEPER'S
BIBLE

DISCOVER CHICKEN BREEDS, BEHAVIOR, COOPS, EGGS, AND MORE

Published in North America in 2023 by Abrams, an imprint of ABRAMS.

First published in the United Kingdom in 2023 by William Collins, an imprint of
HarperCollinsPublishers

Written by Jessica Ford, Sonya Patel Ellis, and Rachel Federman

Library of Congress Control Number: 2022938880

ISBN 978-1-41976413-4

Original illustrations © Lynn Hatzius, see page 413 for details
Design by Eleanor Ridsdale Colussi
Picture research by Milena Harrison-Gray
Chicken-friendly projects by Ryan Ford and Jessica Ford

Printed and bound in Malaysia
10 9 8 7 6 5 4 3 2 1

Abrams books are available at special discounts when purchased in quantity for premiums
and promotions as well as fundraising or educational use. Special editions can also be created
to specification. For details, contact specialsales@abramsbooks.com or the address below.

Abrams® is a registered trademark of Harry N. Abrams, Inc.

DISCLAIMER
The publishers urge readers to be responsible chicken keepers. Please see pages 18–21 for
details on the rules and etiquette of chicken keeping. Additionally, please familiarize yourself
with local restrictions and safety measures.

ABRAMS The Art of Books
195 Broadway, New York, NY 10007
abramsbooks.com

THE HISTORY OF CHICKENS

FROM LEGHORNS TO SILKIES, IDENTIFY CHICKEN BREEDS,

THE
BACKYARD

CHICKEN
KEEPER'S
BIBLE

DISCOVER CHICKEN BREEDS, BEHAVIOR, COOPS, EGGS, AND MORE

DIY PROJECTS CHICKEN COOPS & RUNS

CHOOSING AND STARTING YOUR OWN HOME FLOCK

ABRAMS, NEW YORK

CONTENTS

INTRODUCTION

The world of backyard chicken keeping is rapidly gaining favor as a callback to a simpler life—one that is closer to the land and to the way our food is produced. Thought of in generations past as suitable for farms only, small-scale chicken keeping is now going mainstream. Beyond the allure of fresh eggs, caring for a flock of chickens offers the peace found from interacting with nature, and the joy that comes from raising such beautiful creatures.

Even those who are unable to raise a flock of their own will discover, from perusing the wonderful photos and descriptions in this book, the extent to which these endearing birds can inspire and warm hearts. Learning about chickens normalizes the pastime of keeping them, too, enabling backyard flocks to gain increasing acceptance in our communities.

Chicken keepers are as diverse as the breeds of chickens that they have in their care. Beyond multigenerational farmers and homesteaders, chicken keepers today include

retirees, young families, and even business executives. Motivations and environment vary widely, proving that all can enjoy their own flock, along with the many benefits that these birds provide.

The traits that made the chicken so adaptable to domestication thousands of years ago are what make the modern bird so compatible with backyard keeping today. Very few animals are able to adapt so well to a life in captivity, in so many different environments, while also catering for such a wide range of needs. These modestly sized creatures are affordable to keep, and their independent, agreeable nature means they require very little attention, yet are still content to cuddle on laps. They're also hardy animals, and happiest when living outside with minimal input from their

Above The rich variety of chicken breeds means there is something for every keeper, from cuddly and affectionate pets to egg layers and broilers. **Opposite** This Easter Egger is known as a fancy egg breed, and this breed may lay varying shades of white, pink, brown, green, or blue eggs.

caretakers—so long as they have a safe structure to shelter them from the elements. For many households, these reasons are enough to make the chicken a perfect family pet. For those looking for more, chickens are a reliable source of fresh, nutritious food in the form of eggs and—for those interested—humanely raised meat. With so much to love, it's no wonder that chicken keeping is such a fast-growing hobby among urban dwellers.

The Backyard Chicken Keeper's Bible is the ideal companion for those interested in joining this revival. It's aimed largely at beginner or aspiring chicken keepers, but it will certainly provide new insights and information for more established keepers, too. Establishing a flock can seem daunting at first, but this book aims to clarify what chickens truly need, along with a few things they don't. Aspiring keepers will find that most of the difficulties involved in caring for chickens come at the preparation stage; once you're established, chickens are in fact among the easiest of animals to care for.

Chickens are truly remarkable. They are far more intelligent and complex than most people realize, and their relationship with humans throughout history has often gone far beyond the role of a simple farm animal. For millennia, we have been drawn to the beauty and the diversity of these birds, as well as their unique behaviors. In this book we explore this rich history, together with detailed descriptions and illustrations of the chicken's role in art and culture around the world—from depictions on ancient coins and mosaics, to their symbolism in various religious ceremonies.

A wide variety of appearances is one of the chicken's most noteworthy attributes. There are hundreds of documented breeds in the world, some of which have bloodlines that date back thousands of years, and they

vary greatly in size, color, behavior, and even plumage texture. If you know what it is you need, you will certainly find a breed to suit you. Chapter 1 contains fifty-five of the most desirable breeds in the world today. Carefully selected to include those breeds most popular with backyard and beginner keepers in particular, along with some of the most unique breeds for more specific uses, these profiles alone highlight the surprising diversity of the species.

The past few years have been fraught with uncertainty and isolation. Rumors of food shortages, inflation, and other frightening predictions have left many of us feeling stressed and helpless. We are increasingly seeking relief and a deeper sense of belonging—and remarkable as it may seem, some have found that they can satisfy both of these needs through the simple act of keeping a small chicken flock. Once considered mainly a source of food, chickens are now providing modern keepers with restoration and a stronger sense of connection in uncertain times. By starting a little chicken farm—if only to enjoy the company of these sweet-natured birds—we can improve the lives of our families and our communities.

Right and opposite Chicken keeping provides sustenance in the form of eggs and humanely raised meat, but it also offers many keepers a greater sense of mental well-being, companionship, and community-building.

HOW TO USE THIS BOOK

The Backyard Chicken Keeper's Bible is a tribute to humanity's greatest feathered friend.

Chickens are so much more than an ingredient at the dinner table, as many people around the world are beginning to rediscover. This single domesticated species of bird is incredibly varied in both form and function. Over millennia, chickens have not only been used for food, but also for entertainment, religious ceremonies, fortune-telling, and as status symbols. Today, the chicken is widely associated with a sustainable lifestyle and is even gaining recognition as a family pet. With so much to learn about and celebrate, *The Backyard Chicken Keeper's Bible* is both a practical guide and an homage to this humble bird.

The book begins with Chapter 1: Backyard Chicken Breeds, dedicated to the beauty and diversity of the chicken. Here, we review its fascinating wild origins, as well as its unique anatomy and life cycle. The attributes that set the chicken's wild ancestors apart, from territorial fighting and a complex social hierarchy, to fast-maturing chicks and wide adaptability, are what have made the chicken one of the world's most

populous birds today. Discover a day in the life of a chicken, as well as a stunning array of notable chicken keepers and their birds throughout history and around the world— from Queen Victoria sparking off Hen Fever through to modern-day celebrity chickens and the Chickens of Key West.

Chapter 1 ends with a deep-dive look into some of the dazzling breeds found around the world today. There are fifty-five breed profiles in total, from the widely popular Rhode Island Reds, Jersey Giants, Barred Rocks, and Leghorns, to the lesser-known Sebrights, Langshans, Sapphire Gems, and Egyptian Fayoumis. Each profile features an illustration and short description, along with a quick-view fact file referencing important information such as size, purpose, temperament, and egg-laying ability. Chicken enthusiasts of all kinds will enjoy perusing the breeds listed, and may even find a new breed to fit their interests.

Chapter 2: Chicken Keeping for Beginners offers inspiration as well as a practical guide to getting started. It begins

Each breed is introduced by its name (including any variations), followed by its purpose (the most widely accepted "use" of the particular breed), such as dual purpose breeds, production layers, fancy egg breeds, meat birds (broilers), and bantams and ornamental breeds.

Dual Purpose Breeds
Brahma

The main text for each species provides detailed information on appearance and plumage for all relevant ages and sexes. In addition, behavioral habits and background information are described where it helps the reader gain an insight into the bird in question.

The majestic Brahma is an Asiatic breed believed to have originated in China. It was first introduced to England in the early 1800s, where it was considered so striking and desirable that it helped launch the infamous "Hen Fever" craze (see page 38). It has since contributed to the development of many breeds. Well into the twentieth century, breeders developed several color varieties of Brahma, and these now include the Light, Dark, Buff, Silver, Partridge, Blue, and Lavender. All color varieties feature heavily feathered legs and feet, and small pea combs, making them one of the most cold-hardy chicken breeds available. Brahmas are the "gentle giants" of the chicken world, even towering over other heavy breeds. Roosters reach an impressive 12 pounds (5.4 kilograms), and females may reach 9.5 pounds (4.3 kilograms) or more. Despite its size and beauty, this is a utilitarian dual-purpose breed. Brahma meat is very tender, though birds take nine months to reach full size. They are great egg layers too, even through winter. Hens go broody often and are very good mothers. Brahmas are very gentle, quiet, and laid-back, making them compatible with children, though their size can be intimidating. Roosters tend to be nonaggressive, but they're an imposing foe for anything they deem a threat. Brahmas thrive in smaller spaces, so are a fantastic choice for novice keepers and small backyards.

FACTFILE
Purpose: Dual purpose
Rarity: Uncommon; some varieties rare
Egg production: Good
Egg color: Brown
Origin: China
Size: Heavy
Temperament: Gentle, quiet

The Factfile section provides an easy reference to important information about each breed: purpose; rarity; egg production; egg color; origin; size; and temperament.

59

Opposite Today, chickens are gaining traction as a beloved outdoor family pet.

with a look into the many benefits of keeping chickens in a backyard, including food sustainability, companionship, and building community, and celebrates the renewed enthusiasm for family chicken flocks. This chapter also reviews the fascinating social lives of chickens, and how their daily egg-laying routine works. And for keepers just starting out, a section entitled Hens and Roosters: The Birds and the Bees debunks many chicken myths. For example—did you know that hens don't need a rooster to lay eggs?

Establishing a happy and healthy flock of chickens begins long before the first fluffy chicks are ever brought home. For those unsure of where to start, the section Starting a Home Flock is a helpful resource. It, along with Chicken-keeping Equipment, outlines the major topics that an urban or suburban chicken keeper needs to know in order to set up their own flock for great success. From brooders and feeders, to ordinances and neighborhood etiquette, these sections will help guide beginners toward their first chicken-keeping experience.

Keeping chickens can be a surprisingly complex hobby, and there are many ways to

do it "right." Those with a new flock must make very important decisions early on, such as balancing flock safety with freedom, choosing what feed to buy, and how best to keep their birds safe and healthy during harsh weather. These topics and more are covered in this chapter, with every effort made to ensure readers can make informed decisions on how to care for their flock.

As many chicken keepers will discover, chicken veterinarians are few and far between, so much of a chicken's healthcare must come from its keeper. That's why Chapter 2 contains a vital section on Health and Support, covering the most common injuries and illnesses that a flock may experience, and how best to prevent and treat them. Feather mites, bumblefoot, and egg binding are included here, as are the more serious avian flu, Merek's disease, and salpingitis. Zoonotic diseases—illnesses that may be transmitted from chickens to humans—are covered as well, so keepers know how to keep themselves as well as their birds healthy.

Very few keepers of chickens do so without also tending a garden in their backyard. Finding out how chickens and gardens coexist is an important topic, and one that all gardeners must address if they want their prized garden beds to remain intact. Fortunately, chickens and gardens can coexist—even enhance one another very well—as long as some careful planning is undertaken. This harmonious balance between chicken and garden is covered in Chapter 3: Chicken-friendly Projects, which also includes some helpful hints, and a list of both beneficial and toxic plants.

Chapter 3 also contains a section for DIY-ers who wish to enhance their chickens' lives with a fun project or two. For those with limited tools or experience, Make a Chicken Swing and Make a Chicken Play Gym are great beginner-friendly projects that turn into hours of enrichment for chickens. The more involved Make a Chicken Coop and Make a Chicken Run projects were specially designed for this book. Both are customizable for different sizes and spaces, and have all the features that a flock needs, including roosting bars, ventilation, and predator-proofing.

While at first glance they may seem to have a rather humble appearance, chickens in fact have been a beloved muse and subject for artists for millennia across many cultures. Chapter 4: Chickens in Art highlights some of these beautiful depictions through the centuries, celebrating the gorgeous variety of—and love for—chickens, through the fields of Ornithological Art and Illustration, Painting and Sculpture, Photography and Film, and Design, Craft, and Style. Wherever the domesticated chicken roamed, so chickens in art followed—from eleventh-

century paintings by the Chinese artist Wang Ning to boldly contemporary portraits by Tim Flach, and from the oldest-surviving photograph (1826) of a chicken to Ai Weiwei's bronze rooster head, which has toured the world with its eleven other zodiac animal companions. Featuring Benin Bronzes and Oaxacan *alebrijes*, alongside works by an incredible range of artists across a broad spectrum of media, including Marc Chagall, Pablo Picasso, Gustav Klimt, Fabergé, Ed Ruscha, Francis Barlow, Tamara Staples, and Arthur Parkinson, this chapter will inspire readers to learn more about familiar artists and to discover spectacular new ones, too.

Opposite and below Chickens that are given good care, a healthy and stimulating environment, and plenty of fresh water and food will provide a regular supply of fresh eggs (with egg production dependent on breed).

Following spread For keepers looking for easygoing birds with good egg production in a wide range of colors, look no further than the Easter Egger, seen here enjoying her free-range stroll.

THE RULES AND ETIQUETTE
OF CHICKEN KEEPING

When done right, backyard chicken keeping—even the raising of meat birds—
is more humane than keeping chickens in a factory-farm setting. A well-kept
backyard flock will enjoy ready access to the outdoors, fresh forage to graze on,
ample room to dig and dust-bathe, and a warm, safe place to roost at night.
A poorly kept backyard flock, however, will experience the same health and
well-being issues experienced by factory-farmed birds. The environment that
chickens are kept in is entirely determined by their owner, so it's crucial that it is
appropriate and as humane as possible, whatever the reason for keeping them.

The Five Freedoms

Standards relating to the proper care of
livestock and other animals in captivity
are well established. One particular set of
standards, developed by the Royal Society
for the Prevention of Cruelty to Animals
(RSPCA)—the largest animal-welfare
charity in the UK—is known as the
"Five Freedoms." These offer a good place
to start for chicken keepers anywhere in the
world who want to ensure their flock is
raised ethically and humanely. More can be
found on the RSPCA's website, but in
summary the Five Freedoms are:

- Freedom from hunger and thirst: by
  ready access to fresh water and a diet
  to maintain full health and vigor.

- Freedom from discomfort: by
  providing an appropriate environment
  including shelter and a comfortable
  resting area.

- Freedom from pain, injury, or disease:
  by prevention through rapid diagnosis
  and treatment.

- Freedom to express normal behavior:
  by providing sufficient space, proper
  facilities, and company of the animal's
  own kind.

- Freedom from fear and distress: by
  ensuring conditions and treatment
  that avoid mental suffering.

Left Whatever the purpose for keeping chickens, it is
essential that they are kept appropriately and humanely.

Chicken-keeping rules

Keeping chickens may be more widely accepted now than ever before, but there are still strict rules governing the practice, so it's extremely important to get to grips with any local laws, ordinances, and covenants before you begin. Laws relating to domestic chickens vary widely—from regulations at the federal level, which differ between countries, down to state laws, city ordinances, or local council laws, and even laws established by individual homeowner associations (HOAs). It's up to you as the chicken keeper to fully understand these laws and rules and to abide by them to avoid moving or losing your flock. While rules may change at any time, fortunately most changes tend to be for the benefit of chicken keepers, thanks to a growing acceptance of these birds as pets.

In the United States, all fifty states follow a Right to Farm law (with some variance between states), which protects farmers and homesteaders in rural areas, allowing them to keep poultry and other livestock, as long as they abide by animal welfare and other laws. Unfortunately, the Right to Farm does not extend to those living in urban and suburban areas, but this does not necessarily mean that keeping an urban or suburban flock is prohibited. If you're considering keeping domestic chickens, check out your local city council's website for official rules and regulations. For those unsure where to start, the website TheCityChicken (thecitychicken.com/chicken-laws), launched in 2021, gathers some city regulations for each state in one simple place. In the UK, only flocks of more than fifty must be registered (information is available at www.gov.uk), but local councils and housing authorities will also have their own restrictions. For these—and for guidance in other countries—various online resources will provide all you need to know.

While local laws regarding chickens will look very different between locations, many of them will include the following key points:

- Many cities prohibit certain poultry species like quail and peafowl, allowing only chickens and occasionally ducks.

- Most ordinances allow only a certain number of chickens to be kept on a single property. This limit may range from as few as three adult hens to more than ten.

- Roosters are nearly always prohibited within city limits, due to concerns about noise, though there are sometimes exceptions.

- Like all domestic animals, chickens are not allowed to wander beyond your property. In some locations, they must be kept secured at all times in a designated enclosure.

- Many cities require chicken keepers to purchase permits, which must be renewed annually. In addition, erecting a coop structure on your property may also require a permit.

- Most cities allowing chickens have rules relating to coop size and placement—under a certain height, placed a certain distance from neighboring homes and fence lines, and so on.

- While raising chickens for meat is allowed in some cities, nearly all residential areas prohibit processing them on the property.

- Similarly, there are often strict regulations around selling backyard chicken eggs, even locally.

In addition to the above, covenants and homeowner association rules may determine the appearance of your property, or how it's used—often to the detriment of chicken keeping—and individual lease agreements for renters may prohibit keeping chickens. These regulations can often be very ambiguous, and even contradictory. When in doubt, it's worth considering a visit to a lawyer who specializes in city ordinances and zoning laws. Local policies on chicken keeping may surprise you. In all five boroughs of New York City, for example, there is no limit to the number of chickens that may be kept on a property. However, noise, smell, and other nuisance complaints are taken very seriously. On the other hand, in Florida's Key West, where wild chickens have roamed free since the 1860s (see page 45), regulations are moving in the other direction. In 2004, the city implemented a local chicken catcher, and in recent years, feeding these wild chickens has been made illegal.

Chicken-keeping etiquette

Following local laws is important, but so is maintaining good relations with your neighbors. Fortunately, most people don't mind having backyard chickens in the neighborhood, especially when free eggs are offered! A little consideration will go a long way, though. If you're planning to start a flock of chickens in a neighborhood where people live relatively close together, consider the following:

• Become familiar with any local ordinances, HOAs, and covenant rules, and make sure to always follow them, just in case.

• Avoid placing your coop close to a fence line or a neighbor's house.

• Always keep the coop and run clean. This not only keeps strong odors to a minimum, it also keeps flies under control.

• Construct an inconspicuous or aesthetically pleasing coop and run. Consider keeping the coop height below the fence line, too, or paint it to match your home or fence. Quality construction not only looks nicer, it's also more secure for the chickens.

• Offer free eggs. Many experienced keepers find that this is a very effective way to maintain good relations with their neighbors.

• Only keep breeds known for being quieter and more docile, such as Silkies or Orpingtons.

• Construct a roomy chicken run, and provide plenty of activities. Bored chickens are loud chickens! The more occupied they are with foraging, dust bathing, and roosting, the less likely they are to make a lot of noise.

• Consider a small flock to start. In very densely populated neighborhoods, a flock of five chickens or fewer can be far less intrusive than a flock of fifteen chickens.

• Keep the chickens contained on your property. Allowing them to wander into neighboring backyards is not only dangerous for them, it's a nuisance for the neighbors. Make sure that your chickens cannot easily escape your yard by keeping them in a roofed run, or by ensuring your backyard fence is at least 6 feet (1.8 meters) high. Avoid really flighty breeds, like Leghorns or Polish.

• Consider keeping bantams. Less than half the size of heavy standard breeds, bantams take up far less space, and they tend to have quieter voices. Some breeds can be decent layers, too, especially for a small family.

Opposite When planning a new flock, be sure to consider local laws as well as the physical space. If you live in an area with nearby neighbors, consider keeping five chickens or less, or even smaller, quieter bantams.

A BRIEF GLOSSARY

Chicken anatomy and chicken keeping come with their own jargon—a set of widely accepted terms used when referring to chickens, their housing, or their feed. Since these terms appear frequently throughout this book, readers are encouraged to familiarize themselves with them to better understand the material.

Autosex: A rare trait found in some breeds that allows all chicks, regardless of generation, to be easily sexed between male and female at hatching.

Bantam: A miniature-sized chicken breed. Usually weighing 32 ounces (900 grams) or less. Generally kept for show or for companionship.

Bedding: Material used to cover the floor of the coop. Usually comprised of sand, straw, or pine shavings.

Broiler: A chicken breed solely developed for fast, efficient meat production.

Chick: A baby chicken, from hatching age to about six weeks, when fully feathered.

Chick starter/grower: Chicken feed formulated for chicks, pullets, and cockerels. Also fed to non-laying hens and roosters.

Cock: A colloquial term for "rooster." *See* Rooster.

Cockerel: A young male chicken. Fully feathered and possibly crowing, but not yet at sexual maturity.

Comb: A fleshy growth on top of a chicken's head, starting just above the beak.

Coop: An enclosed structure designed to house chickens, particularly at night.

Dual-purpose: A chicken breed developed for both meat and egg production.

Earlobes: A fleshy patch of skin located just below the ear canal.

Embryo: A viable chick not yet hatched from an egg.

Flock raiser: Feed for chickens, pheasants, and other domesticated fowl. Formulated for fowl of all ages.

Forage: Mixed leaves, grasses, shoots, and bugs, as found in a typical field for chickens to eat.

Free-ranging: The act of letting chickens wander free in a large, open area.

Hackle feathers: Feathers on the back of the neck. More pronounced in roosters.

Hen: An adult female chicken that has begun laying eggs.

Hybrid: A designer chicken that is the offspring of two separate chicken breeds.

Layer: A chicken breed developed for high egg production.

Layer feed: Chicken feed formulated only for laying hens.

Nest box: A dedicated space, about 12 x 12 inches (30 x 30 centimeters), provided for hens to lay their eggs.

Pin feathers: The initial feather growth as it emerges from the skin. The growth is hard to the touch and ends in a sharp tip before the soft filaments emerge.

Pullet: A young female chicken. Fully feathered, but not yet laying eggs.

Roost: A bar or branch for chickens to perch on when sleeping at night.

Rooster: An adult male chicken who has begun to mate.

Run or pen: An open, fenced-in enclosure designed to keep chickens contained during the day.

Saddle feathers: Feathers on the back of a chicken, just above the tail. More pronounced in roosters.

Scratch: Chicken feed made of mixed grains. Fed as a treat, not as a main food.

Sex link: A hybrid breed that allows the first generation of chicks to be easily sexed between male and female at hatching.

Sickle feathers: Long, decorative feathers on the sides of the tail. Only found in roosters.

Spurs: A horn-like growth on the back of a chicken's legs. Develops into a long spike in roosters.

Wattles: Two fleshy growths below a chicken's chin, starting just behind the beak.

CHAPTER ONE

BACKYARD CHICKEN BREEDS

The origin of chickens

Tracing the origins of domesticated chickens (*Gallus gallus domesticus*) lies at the heart of ongoing research involving a rather complex study of mitochondrial DNA. (And separately, but not totally unrelated, American paleontologist Jack Horner is currently trying to reconstruct a dinosaur using chicken DNA.) Today's chickens likely came from the Red Junglefowl (*Gallus gallus*) native to the jungles of Southeast Asia. Some sources also point to the Gray Junglefowl (*Gallus sonneratii*) of south India as an ancestor. Were chickens domesticated in two distinct places? Or did a domesticated version "fly the coop" and mix with a wild one? These remain open questions that scientists continue to investigate.

We know chickens aren't big travelers and their flying abilities are limited. These made them good candidates for domesticity and breeding. But how did this originally take place? Centuries of potential mixing between domesticated and wild chickens make tracing their DNA history even more challenging.

Evidence suggests that domestication of these now ubiquitous creatures began in the range of 8,000 to 10,000 years ago (see also Chinese and Japanese Art, page 280). Fossils believed to be chicken bones were

found in northern China, in what is now Hubei province, thereby offering up a potential site of early chicken domestication. The Indus Valley is also an important place on the map of chicken history. Chickens—or their ancestors—were even mentioned on cuneiform tablets from Mesopotamia (present-day Iraq): the Sumerians called them "the bird of Meluha," Meluha being the regional name for the Indus Valley.

It is likely that chickens made their way from Mesopotamia to ancient Egypt, where they were employed in fighting for entertainment and kept as exotic pets. The annals of the pharaoh Thutmose III indicate that chickens were a novelty during his reign, called "the bird that gives birth every day." It took about a millennium for the Egyptians to really get their egg-laying methods off the ground, using organized incubation. The ancient Greek historian Diodorus Siculus wrote about the impressive efforts in his *Library of History*. Mud ovens recreated brooding conditions with heat from a nearby controlled fire, leading to enhanced production. Later, across the Mediterranean, Romans enjoyed chickens in feasts and even imposed limits on consumption and on the practice of artificially fattening the fowl in an early nod to humane animal husbandry.

As for chickens in the "New World," it's possible the Polynesians brought chickens to South America in the fourteenth century. Some say they first arrived on Columbus's second voyage. Long before it was a USDA standard, chickens in the Americas lived free-range lives and were primarily kept for their eggs. In the 1700s chickens were a popular sight in the colonies, found wandering on farms as well as city streets. In 1986, Colonial Williamsburg established its Rare Breeds program, featuring (and protecting) breeds that were common in the eighteenth century but are now threatened or possibly endangered. Dominiques and Nankin Bantams are two of the breeds that delight visitors at the historical village today.

Before Hen Fever exploded in the mid-nineteenth century, President Thomas Jefferson's home at Monticello hosted a bounty of Bantams, which seemed to offer their owners a great deal of pleasure. In the fall of 1806 Jefferson wrote to his granddaughter, "… I send you a pair of bantam fowls;

Above Chickens, known as "the bird that gives birth every day," appeared on 11th-century tomb walls in ancient Egypt.

quite young: so that I am in hopes you will now be enabled to raise some. I propose on their subject a question of natural history for your enquiry: that is whether this is the Gallina Adrianica, or Adria, the Adsatck cock of Aristotle? For this you must examine Buffon etc." The mid- to late nineteenth century then saw the hen truly trend, with an intensity of interest in breeding, publishing, and exhibition.

Still, at the start of the twentieth century, chicken meat and eggs were not everyday foods; on many farms there was not a cluck cluck here, there, or anywhere. Enter World War I, when the Daughters of the American Revolution hatched a plan to help France replenish their fresh food. The group began to fundraise for eggs and chickens specifically. (Or *oeufs et poules*, as they would be known when they were sent overseas.) Homesteading podcasters and bloggers might encourage keeping backyard chickens today, but in 1918 the US government practically demanded it with a poster by the USDA announcing: "Uncle Sam Expects You to Keep Hens and Raise Chickens." Rationing during World War II led to another request for and surge in chicken keeping as part of the "victory garden" effort.

By mid-century, the sustainable practice of backyard chicken keeping was on the decline. Fortified feed replaced the need for sunlight, the birds were moved to tight, windowless quarters, and true factory chicken farms were perfected. Selective breeding led to a deliberate divergent path, depending on the end game: layers are chickens kept for their eggs, and broilers are chickens reared for their meat. The bird that once announced the dawn now spends its life without seeing it.

With the dawn of a new century, backyard chicken keeping is once again on the rise. As chickens become increasingly common in suburban backyards and even cities, it seems safe to say that the ever-evolving fowl story that began in the jungles of South Asia with a low-flying, slow-moving bird isn't over.

Top Chickens, or their ancestors, were mentioned on cuneiform tablets from Mesopotamia (5–6,000 years old). **Above** Chicken keeping was deemed a national duty during both World Wars, including in the US, France, and the UK. This 1916 poster displays the significance of chickens to the war effort. **Following spread** Two single-comb white Leghorn cockerels on exhibition at the New York Poultry Show in 1933, with more than "3,000 feathered aristocrats as attractions."

Chicken anatomy

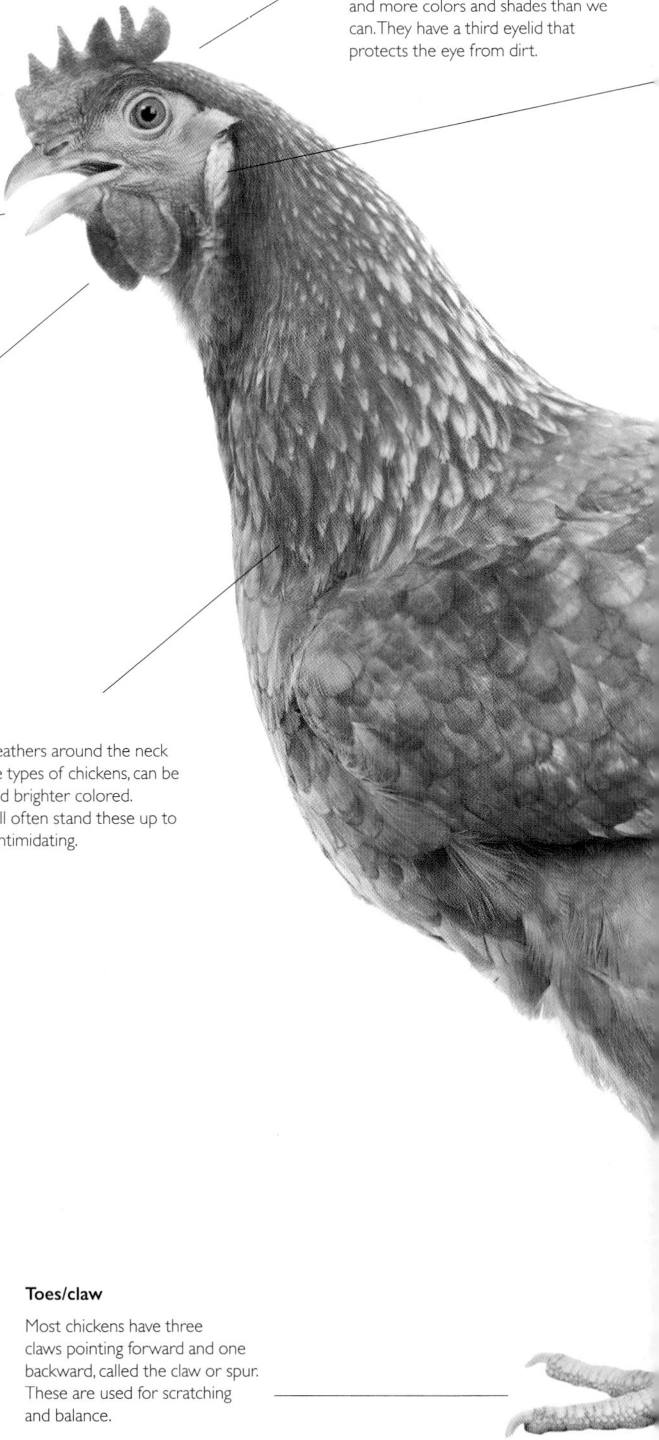

Comb

The comb is a fleshy crest on top of the head. It is actually an organ used to regulate body temperature. A healthy chicken's comb will usually be bright and red.

Beak

Not only is the beak for eating, it is also used to regulate body temperature by "panting."

Eyes

Chicken's eyes can see in 300 degrees and more colors and shades than we can. They have a third eyelid that protects the eye from dirt.

Wattles

Wattles are fleshy skin flaps that hang on each side of the throat. Blood circulates from the comb to the wattles to help keep the chicken cool.

Hackles

These are feathers around the neck and, in some types of chickens, can be long, fine, and brighter colored. Roosters will often stand these up to look more intimidating.

Toes/claw

Most chickens have three claws pointing forward and one backward, called the claw or spur. These are used for scratching and balance.

Ear lobes

Ear lobes are located inside the head but are visible on the outside. There are exceptions, but often chickens with white ear lobes tend to lay white eggs; chickens with red lobes will lay brown.

Main tail

Roosters' tails are often larger plumages; hens are smaller.

Vent/cloaca

This single opening is responsible for peeing, pooping, and egg laying.

Shank

Also known as the leg, chicken shanks can be feathered or simply scaled. Most are yellow or white, but some can be green or black.

A day in the life of a chicken

Chickens are early birds. They'll usually wake with the sun, starting their day by 7 A.M. at the latest. They need to make use of the sunlight as they have no ability to see in the dark, and they also require that vitamin D to lay eggs—one of their top priorities on any given day. They're ready to eat soon after stretching their legs, so once they're moving about in the fresh air, they'll appreciate fresh water and a food refill. They'll eat scratch (wheat and corn) and, strange as it sounds, grit in the form of tiny pebbles (also produced commercially, made from either crushed shells or flint), which helps them grind up food in their gizzards. Chickens are also big fans of table scraps, happily eating lettuce, eggshells, fruit and vegetable peels, and most other leftovers. Drinking is a slow process for a chicken. They pick up a little liquid and tilt their head back so the water can travel down their throat. (See pages 210–13 for more on water and feed.)

After the morning meal, hens are likely to move on to egg-laying. If you're nearby, you might overhear a cackling song during this time. Farmers generally come through to collect the eggs shortly after the chickens are done laying them in the morning, and at least once more later in the morning or in the afternoon. Hens will start laying eggs at six months—on average two eggs every three days. In the first year, a hen will lay about 240 eggs in total. If they're not collected, she will sit on groups of eggs (called a clutch) for three weeks. By eighteen months she's through with her greatest egg-producing years, which come to a complete end by three or four years. If kept safe, a chicken can live up to ten years or so.

Some people let their chickens run around the yard, especially if they're able to keep an eye on them. The exercise is good for them, as is the chance to find worms and insects. Otherwise, they'll enjoy getting some exercise in a run with a fence around it. Predators include dogs, cats, racoons, opossums, foxes, weasels, hawks, and more; with so many enemies, it's no wonder chickens can be a bit … chicken sometimes. During the day they'll manage to stay out of harm's way, but because they can't see at night, they must have a sheltered place to sleep (see pages 198–205).

Chickens are gentle and enjoy being around humans. Their owners can pick them up and feed them by hand. Some hens like to be petted. Other daily activities include preening, taking dust baths (see page 184), and singing little chirping songs.

If there are chicks around, mother hens will take good care of them, helping them find food in a process called tidbitting, where they point out "tidbits," teaching them to learn by trial and error what makes a good snack. During courtship, roosters tidbit as well, pointing out tasty items on the ground to a potential mate. Courtship also involves performing a little dance, placing one wing on the ground and prancing around a bit (see page 175). In addition to courting the ladies, roosters keep vigil over the flock.

Besides all this, you'll see a lot of pecking and scratching at the ground—chickens enjoy the process of foraging for food. And a chicken wiping its beak on the ground is cleaning and sharpening it, getting it ready for the next round of pecking.

Chickens call each other when it's time for bed.

Chickens know when it's time to head to bed, and they're happy to settle in for the evening. The inside of their coop might be covered with wood shavings, hay, or even sand, and the perch or roost—consisting of a bar, branch, or a piece of wood—is where they'll sleep, holding on and standing up.

A rooster crows. A new day begins. But the rooster isn't welcoming the day so much as marking territorial possession. Yet for many cultures, these regal creatures have symbolized hope and a new start. So, are roosters waiting for those first rays of light to make their way over the horizon? A 2013 study at Nagoya University in Japan showed that roosters rely on an internal clock to signal the start of the day. Roosters also let out their famous cock-a-doodle-doo to warn of predators or to draw attention to an appealing snack.

In the past decade, there has been a move toward keeping "free-range" chickens, but the majority of chickens no longer soak up the sun or peck around a yard. Most live in factories, with tens of thousands packed closely together, with no access to the sun.

Left and below While factory chickens are still commonplace around the world, there has been a recent and widespread surge of interest in free-range chickens, whether for eggs, meat, or companionship in small home flocks. **Following spread** Rhode Island Reds can make great mothers.

A CHICKEN'S LIFE CYCLE

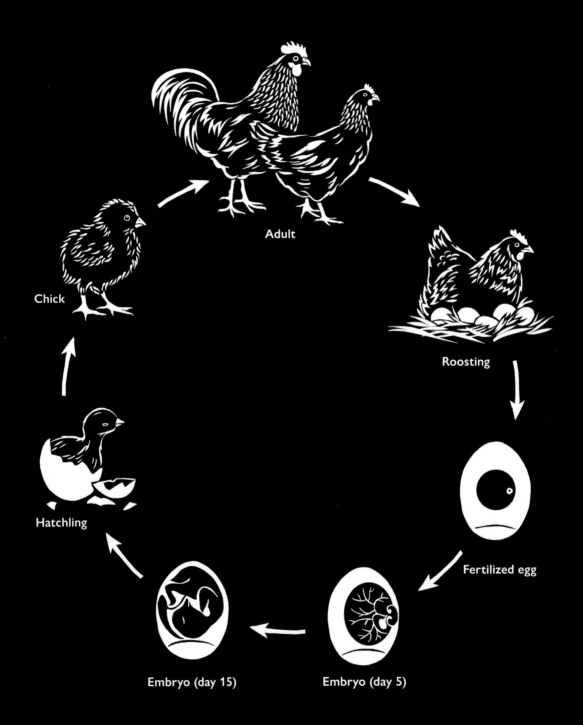

Adult

Roosting

Fertilized egg

Embryo (day 5)

Embryo (day 15)

Hatchling

Chick

- **DNA exchange:** A hen and rooster mate, giving a hen the sperm she needs to create a fertilized egg.

- **The ova grow:** Inside the chicken, each ovum grows a yolk and then the egg white, which encases the yolk. The shell forms last. Sperm from the rooster makes its way into a yolk, fertilizing it.

- **Entering the world, almost:** The hen lays the egg. She lays on average two eggs every three days.

- **The hen goes broody:** After laying about eight to twelve eggs (a clutch), a hen turns her attention to hatching. At this point she will sit on her clutch to keep it warm. This period is also referred to as incubation.

- **The chicks appear:** With the right conditions (staying warm and being turned over) in three weeks the chicks will hatch.

- **The first few weeks:** Young chicks stay with their mothers for warmth, protection, and help finding food. Her clucks lead them to their meals. On farms and in backyards, chicks are often given a high-protein starter feed. In the wild, they'll eat insects or worms.

- **Growing up:** At six weeks, the once-fuzzy chicks have replaced their down with feathers and now look like mini-chickens. A pullet is a young hen; a cockerel is a young rooster.

- **The cycle begins again:** At five months, a pullet can start laying eggs.

- **Full circle:** At one year, the chicks are fully grown members of the flock.

A royal flock and other notable urban chickens

In 1842, not long into her reign, Queen Victoria received two roosters and five hens as a wedding present. Naval Officer Edward Belcher had brought the exotic "Cochins" back with him following his voyage to China, Sumatra, and Vietnam. (For context, the first Opium War had just ended, which had resulted in China ceding Hong Kong to the British Empire.) Victoria and Albert valued the birds for their large size, feathered feet, and bright colors. Chickens had been in Europe for many centuries—across the channel, Marie Antoinette had once hosted a hen house at Versailles—but these gifted birds were three times bigger than the average nineteenth-century hen.

> *"Her Majesty's collection of fowls is very considerable, occupying half-a-dozen very extensive yards, several small fields, and numerous feeding-houses, laying-sheds, hospitals, winter courts, &c. ... The eggs are of a deep mahogany colour, and of a delicious flavour. These birds are very healthy, quiet, attached to home, and in every respect suited to the English climate. They are fed, like most of the other fowls, on a mixture of boiled rice, potatoes, and milk."*
> *Illustrated London News,*
> **December 23, 1848**

On the farm next to Windsor Castle, the "Cochin China Fowl" were given their own gothic-style house. They roosted and nested in the main pavilion and had space for running and pecking around in the yard. So taken was the Queen by her flock that she had an apartment built next to the poultry house. From that perch, she could watch her feathered friends up close (along with a cup of tea, of course). During this period, some chickens still ended up on the dinner table; others were given as gifts. Eggs from the Cochins were carefully wrapped and sent off to friends. And so began a nineteenth-century craze.

"Hen Fever"(see also page 277) or "The Fancy," as the Victorian hen trend was called, soon spread to the New World where it lasted around a decade. Poultry shows came to town, books were published, and hen enthusiasts experimented with all kinds of creative breeding. The Rhode Island Red first appeared at this time in the small New England state of that name and neighboring Massachusetts, the result of cross-breeding breeds from Asia with Italian breeds. And the Dominique, or Pilgrim's Fowl, also derives from this period—believed to be the country's oldest original breed. Boston's Public Garden hosted one of the country's first poultry shows in 1849. Ducks (now famously associated with the public park thanks to the 1941 picture book by Robert McCloskey, *Make Way for Ducklings*), pigeons, and even swans joined the exhibition, but judging was canceled following confusion over what constituted a pure breed. (In 1874, the American Poultry Association introduced the American Standard of Perfection.) The action also moved down the coast—in 1854, circus king P. T. Barnum hosted two poultry shows at his American Museum in New York.

American poet Robert Frost was inspired by another New York exhibition (this one at Madison Square Garden) when he wrote the short story "Dalkins' Little Indulgence—A Christmas Story," which was published in *The Farm-Poultry Semi-Monthly* in 1905. (Frost kept chickens in his Derry, New Hampshire farm, and members of the henhouse showed up in other stories and a poem as well.)

Chickens were getting attention on an even bigger stage around this time. President Teddy Roosevelt's animal-loving family brought their colorful collection to the White House. Among their many beloved pets domestic and wild was a one-legged rooster named Fierce and a hen named Baron Spreckle. (A badger, a macaw, a small bear, a lizard, guinea pigs, a regular pig, a pony named Algonquin, a hyena, a barn owl, and a pack of dogs joined in the fun along with other pets.)

Not long after Robert Frost's poem about a prize-winning hen was published, down in Alabama a young boy named John Lewis cared for and even started preaching to the chickens on his farm. Lewis, who grew up to be a civil rights activist and beloved congressman, told the story of his barnyard friends in his graphic-novel memoir trilogy *March* (Top Shelf Productions, 2013–16). A picture book of his childhood authored by Jabari Asim and illustrated by

PROBABLE RESULT OF THE POULTRY MANIA.

Above "Probable Result of the Poultry Mania" sees chickens bred to gargantuan proportions, from the satirical magazine *Punch*, volume 24, 1853. **Below** This 1906 photograph of a brooder on a chicken ranch in Sonoma, California shows the reach of "Hen Fever."

Clockwise (from top left) Poet Robert Frost included his pet chickens in his poems, US Congressman and activist John Lewis preached to his beloved chickens, President Roosevelt's one-legged rooster "Fierce" lived at the White House, and Circus King P. T. Barnum hosted NYC poultry shows in 1854.

Earl Lewis was also published in 2016, aptly titled *Preaching to Chickens* (Penguin). In a commencement speech at Washington University in the same year, John Lewis talked about the rapt audience he had, which was believed to include Dominiques, bantams, and Rhode Island Reds: "…I would preach to these chickens. And some of these chickens would bow their heads, some of these chickens would shake their heads."

By the twentieth century, chicken had become a staple on dinner tables around the world, but the barnyard fowl have continued to capture the hearts, not just the appetite, of many. Along with her retired NFL quarterback husband Tom Brady, Brazilian model Gisele Bündchen cares for chickens in her Los Angeles garden; the hens eat scraps from the family's vegetables and in return the family gets fresh eggs. Actor Julia Roberts is also known to appreciate how keeping chickens contributes to a sustainable lifestyle. Read on for more celebrity hens on both sides of the pond.

Celebrity chickens and chicken keepers

Nicole Richie
Tallulah, Philomena, Mama Cass, Sunny, Daisy, Ivy, Sibby, and Dixie Chick get the star treatment like their owner—they live in a house modeled after Richie's.

Serena Williams
In 2021, Alexis Ohanian—the tennis champ's husband—posted an image of the family's hens on Instagram. Bum-Bum, Chikaletta, Minnie, and Daisy were named by the couple's then three-year-old daughter.

Elizabeth Hurley
Chickens enjoy the country life on the actress's organic farm in Gloucestershire.

Jennifer Aniston
Season 3 of *Friends* introduced us to Chick and Duck, unusual additions to West Village, New York City life, but beloved pets of Chandler and Joey. In real life, Jennifer Aniston raised chickens with her former husband Justin Theroux and enjoyed fresh eggs from their flock, making carbonara (a pasta from Rome made with eggs, cheese, black pepper, and pancetta) among other egg-centered dishes.

Jennifer Garner
Actor, activist, and cofounder of an organic baby-food line, Jennifer Garner keeps several chickens. She was once spotted walking one of them—Regina George, who was said to love kale—on a leash.

Reese Witherspoon
The leading lady from Louisiana has a penchant for Silkies; they're a little on the smaller size (but the yolks are big!). She keeps her flock of twenty at a ranch in California.

Tori Spelling
The former *90210* star's fashion sensibilities extend to her feathered friends. She apparently used to create outfits for her white Silkie, Coco (named, fittingly, after Coco Chanel), but that's not all. Coco also got a primo spot on Tori's bed. This was a one-of-a-kind,

Left Actor Tori Spelling and Coco, her pet Silkie, wear matching dresses onstage at the 22nd annual GLAAD Media Awards in Los Angeles in 2011.

chicly dressed chicken with her own poncho, but that didn't stop people from thinking they were spotting a poodle in Tori's arms rather than a member of the Phasianidae family.

Isabella Rossellini

Actor, artist, filmmaker, and model Isabella Rossellini has added one more string to her bow—devoted chicken keeper. With names like Andy Warhol and Amelia Earhart, her brood of around 120 heritage chickens live on "Isabella's Farm" in Long Island, where she grows vegetables and produces honey and eggs. Her book, *My Chickens and I* (Abrams, 2018) encompasses photographs by Patrice Casanova, observations, fun facts, and hand-drawn illustrations.

A royal tradition lives on

Queen Victoria would be happy to know the hen hobby lives on among the British royal family. King Charles III and Camilla, Queen Consort, are such big chicken fans that their Gloucestershire home has earned the nickname "Cluckingham Palace." In 2018 the then-Prince was shown feeding his Marans at Highgrove House in an Instagram photo taken by the royal photographer Chris Jackson. In a foreword to *The Illustrated Guide to Chickens: How to Choose Them, How to Keep Them* by Celia Lewis (Bloomsbury, 2010), the King mentions his great-great-great-grandmother Queen Victoria's flock.

Both of the King's children are continuing the tradition with their own families. Princess of Wales Kate Middleton and Prince William enjoy chicken rearing with their children in Norfolk. George, Charlotte, and Louis help collect the eggs (and eat them). Meanwhile, across the pond, the Duke and Duchess of Sussex (Prince Harry and Meghan Markle) proudly showed off the coop they call "Archie's Chick Inn" on an Oprah TV special. The hens were rescues, a passion of Meghan's. (Side note: Oprah herself has been known to enjoy chicken keeping at her Hawaiian home.)

Top This iconic 1867 photograph shows the animal-loving queen with her dog Sharp at Balmoral. Queen Victoria held a central role in the ongoing love for chicken keeping, starting "Hen Fever" (also known as "the Fancy") when she was given two rooster and five hens as a wedding present. **Bottom** 1843 painting shows Queen Victoria at that time. The hen hobby lives on among the current British royal family.

Hawk Mother tells the real-life story of a Red-tailed Hawk named Sunshine who is shot and no longer able to fly, and is taken in by zoologist Kara Hagedorn. Sunshine lays eggs each year, but they turn out to be infertile each time, so Kara decides to experiment by giving her two chicken eggs instead. In the wild, hawks are dangerous predators to chickens, but Sunshine keeps her "adopted" eggs—and later chicks—warm, safe, and well fed until they're ready to fly the coop to a farm. Learn more about Sunshine and Kara at Hawkmother.com or through the pages of their book, *Hawk Mother: The story of a red-tailed hawk who hatched chickens*, winner of the Flora Stieglitz Straus Award for Best Nonfiction Book for Children and selected for both the Junior Library Guild Selection and NSTA/CBC Outstanding Science Trade Book for Children.

Above *Hawk Mother* is available for purchase online, and there are free lesson plans available through Web of Life Children's Books.
Below Sunshine (right) and one of her adopted eggs, now hatched and grown into a full-size chicken named Gaia.

The chickens of Key West

Travelers come to Florida's Key West for its world-famous sunsets, turquoise water, live music, and perhaps a tour of Ernest Hemingway's Spanish Colonial home. But they might be surprised to find feathery revelers among the crowds on Duval Street, giving new meaning to the term "free-range." The chickens' ancestors were Caribbean junglefowl—once an important source of food on the island—and roosters that were unleashed in the 1970s following a prohibition on cockfighting.

Known as the Key West Gypsy Chickens, these lovable locals can be seen safely crossing the street without crossing guards and heard clucking at all hours. At the iconic Blue Heaven, a popular restaurant on Thomas Street, you can enjoy Breakfast with the Roosters, choosing from pancakes, omelets, and other options as you enjoy the company of free-roaming fowl. While tourists tend to be enchanted by these feathery vacation guests, and eager to capture the hens for their Instagram feed, some locals are concerned about the noise, droppings, and threat to native species. An ordinance was passed in winter 2021, making it against the law to feed the fowl, but the chickens are fairly self-sufficient, and their penchant for insects makes them natural allies in the move to organic farming. Non-residents looking for a live souvenir once they leave the palm-tree-lined streets of the Keys can adopt their own chicken at the Key West Wildlife Center. Each chicken comes with a letter signed by the Mayor, certifying it as an authentic Key West Gypsy Chicken, and an agreement must be signed to say that it will be kept as a pet and not for meat.

Opposite, top, and left The Key West Gypsy Chickens have become a tourist attraction of the beautiful Florida Keys in their own right. They can be seen strutting nonchalantly outside the crowded bars and key lime pie purveyors of Duval Street and merrily pecking in the hallowed spaces of the Memorial Sculpture Gardens and Hemingway House.

AT-A-GLANCE
BACKYARD CHICKENS

Dual-purpose / Heritage / Homestead

Orpington

Australorp

Rhode Island Red

Plymouth Rock

Brahma

Buckeye

Sussex

Faverolles

Naked Neck (Turken)

Jersey Giant

Wyandotte

Bielefelder

Dominique

Welsummer

Barnevelder

New Hampshire Red

Dorking

Langshan

Scots Dumpy

Minorca

Fancy Egg Breeds

Easter Egger /
Americana

Araucana

Olive Egger

Ameraucana

Legbar

Marans

Whiting True Blue
& Green

Production Layers

Leghorn

Production Red

Amberlink

Red Star

Black Star

ISA Brown

Sapphire Gem

Meat Birds / Broilers

Cornish Cross

Bantams & Ornamental Breeds

Broiler Hybrids

Old English Game
Bantam

Modern Game Bantam

Cochin

Sebright

Silkie

Sultan

Belgian D'Uccle

Japanese Bantam

Serama

Hamburg

Polish

Ayam Cemani

Egyptian Fayoumi

Yokohama

Onagadori

Dong Tao

Sicilian Buttercup

Campine

Frizzle

BACKYARD CHICKEN
BREED PROFILES

Thanks to the chicken's popularity and dizzying diversity, records of breed descriptions can be incredibly complex. There are several hundred breeds currently recorded, each of which may have dozens of known variations in size and color, and countless new breeds and hybrids are in development every year.

Information on the breeds and hybrids listed in this section reference the American Poultry Association, the Poultry Club of Great Britain, and descriptions of breeds provided by reputable hatcheries. Due to wide regional variation, some breed information presented is intentionally simplified, and may conflict with specific breed registries and clubs.

Each profile includes the following information:

Breed name
The registered breed or hybrid name. For unregistered breeds, the most common name is given.

Purpose
The most widely accepted use of the breed.

Rarity
The scarcity of the breed, based on availability in the United States and United Kingdom.

Egg production
The most accepted range of egg production for the breed, based on average eggs laid per year. Poor is under 100 eggs/year, good is 100 to 200 eggs/year, very good is 200 to 250 eggs/year, and excellent is in excess of 250 eggs/year.

Egg color
The pigment of the outer eggshell of eggs laid by the breed. Actual egg color may vary by individual hen.

Origin
The region in which the breed was formally developed, though its genetics may originate from elsewhere.

Size
The range of adult chicken weight. The actual breed size can vary, depending on location and strain.
Note: most standard breeds have smaller, bantam counterparts that may not be covered in this section.

Temperament
The accepted known temperament of the breed. Actual temperaments will vary widely by individual.

Opposite This Polish Frizzle is a stunning example of an ornamental bantam breed.

DUAL-PURPOSE BREEDS

Dual-purpose chicken breeds were developed all over the world to be the perfect family homestead chickens. These are the tried-and-true chicken breeds, most of which have lived on farms since at least the early twentieth century. Although they don't lay as well as the production hybrids (see page 116), or produce meat as quickly as broilers (see page 136), these chickens produce both high-quality meat and dependable quantities of large eggs. Dual-purpose breeds were developed to be utilitarian birds, able to withstand harsh climates and a life out on the farm. The result is a wide variety of hardy, healthy breeds to choose from, perfect for the backyard flock.

Opposite and above Dual-purpose chickens were bred to be homestead chickens. They are hardy and can survive harsh climates. They come in a wide variety of colors and body shapes.

Dual-purpose Breeds

Orpington

One of the most quintessential heritage breeds, the Orpington is famous for its soft appearance, docile temperament, and dependable egg-laying ability. The breed was originally developed in the town of Orpington, in southeast England, by the famous breeder William Cook. By 1895 Cook had developed a well-rounded homesteading bird, the Black Orpington, which produced great-quality meat and eggs. Just a short time later, the breed arrived in the United States where it was well received. The Orpington remains very popular today. There are several color varieties available, including Buff, White, Black, Blue, Chocolate, and Lavender, with the Buff being by far the most common and beloved—fondly known as the "golden retriever of the chicken world." The hens are good layers of large, brown eggs. They tend to go broody often and are very attentive mothers. The roosters are big and beautiful; they're protective of their hens, but generally not human-aggressive. This is a heavyset breed, with roosters weighing up to 10 pounds (4.5 kilograms), and hens averaging 8 pounds (3.6 kilograms). Orpingtons of all colors are known for their soft, fluffy appearance, and very social nature. They are ideal for beginners and families with children.

FACTFILE
Purpose: Dual purpose
Rarity: Common; some varieties rare
Egg production: Good
Egg color: Brown
Origin: England
Size: Heavy
Temperament: Docile, quiet

Above The original was the Black Orpington (seen here), and several color varieties now exist. **Opposite** Known as the "golden retriever of the chicken world," the most common Buff Orpington variety is ideal for beginners and families with children.

Australorp

Considered by many to be one of the best backyard chickens, the Australorp is a very popular breed, known for its gentle nature and excellent egg-laying ability. As indicated by its name, the Australorp was developed in Australia from the Black Orpington (see page 53). It was developed as a dual-purpose bird for Australian homesteads, and had made its way to the United States by 1920. This is one of the best heritage egg-laying breeds, once holding the record for the most eggs (364) laid in a year. The Australorp is very close in appearance to the original Black Orpington, with glossy black feathers, dark legs, and a bright red comb and face. The roosters, which tend to be less aggressive than most, are especially handsome, with an iridescent green sheen to their feathers. Some other Australorp colors were developed, but they are very rare. The breed averages 8.5 pounds (3.9 kilograms) for roosters, and 6.5 pounds (2.7 kilograms) for hens. Australorps are known for their quiet temperament, laidback personality, and ability to thrive in contained spaces. They are an excellent choice for chicken-keeping beginners, and families with children.

FACTFILE
Purpose: Dual purpose
Rarity: Common; some color varieties very rare
Egg production: Excellent
Egg color: Brown
Origin: Australia
Size: Heavy
Temperament: Friendly, quiet

Above Developed from the Black Orpington, the Australorp is also considered to be one of the most popular choices for novice chicken keepers. **Right** Australorps have a soft green iridescent sheen to their black feathers, which is beautifully contrasted by their bright-red comb. **Following spread** These Australorp rooster's feathers display this green iridescence even more prominently.

Dual-purpose Breeds

Rhode Island Red

The Rhode Island Red is an all-American heritage breed. Since its development in the late nineteenth century, it has been one of the most popular dual-purpose breeds, thanks to its hardiness and excellent laying ability. This breed originated somewhere between Narragansett Bay and Buzzard's Bay, and was the result of crossing Leghorns with Asiatic breeds like Cochins (see page 149). The Rhode Island Red has diversified in recent years, with many egg-laying hybrids and production strains existing alongside the more traditional bloodlines. Production strains and hybrids tend to be much lighter in color, and egg production is higher. The traditional Rhode Island Red is a striking dark mahogany color, with a yellow beak and legs. Roosters sport a large, bright red comb, and shiny black feathers in the tail. This is a heavy breed, with roosters weighing in at 8.5 pounds (3.9 kilograms), and the hens at 6.5 pounds (2.7 kilograms). Very active and intelligent, the Rhode Island Red forages well and is a great free-ranging backyard chicken. Some members of this breed may have a tendency toward aggression—especially the roosters—but the hens can also be very friendly and gentle.

FACTFILE
Purpose: Dual purpose
Rarity: Common
Egg production: Excellent
Egg color: Brown
Origin: United States
Size: Heavy
Temperament: Varies, but generally active, intelligent, and friendly

Above The standard Rhode Island Red has bright-yellow legs and beak, set off against the deep mahogany of its body. **Left** Rhode Island Reds are great all rounders, making them excellent for novice and experienced chicken keepers alike. They are also some of the friendliest chickens and make great pets.

Dual-purpose Breeds

Plymouth Rock

The Plymouth Rock is another all-American breed, developed in the early nineteenth century. This breed comes in several colors, including Buff, Partridge, White, and Barred. The Barred Plymouth Rock is by far the most famous, usually going by the name "Barred Rock." It's distinguished by its namesake narrow barring of dark gray and white, with yellow legs and a bright red comb and face. The Buff, Partridge, and White Rocks are less common, but they share the same deep-bodied build. No matter the color, this breed is one of the most dependable dual-purpose heritage breeds available—so much so that it was used for the development of several broiler and egg-production hybrids. Today, the Plymouth Rock remains among the top ten backyard chicken breeds. It is cold-hardy, friendly, and an excellent choice for beginners and families with children.

FACTFILE

Purpose: Dual purpose
Rarity: Common; some varieties uncommon
Egg production: Very good
Egg color: Brown
Origin: United States
Size: Heavy
Temperament: Quiet, friendly

Above While there are several color varieties, the barred coloration (seen here) is the most famous, with stripes of black and white. Barred Rocks are a wonderful combination of hardy chickens with stunning looks. **Opposite top** Before they show their iconic barred markings, Barred Rock chicks are dark gray or black with occasional yellow patches. **Opposite bottom** Plymouth Rocks are generally docile and can get along well with other chickens. However, if unsupervised, they can occasionally be a little pushy, and many chicken keepers suggest grouping them with Faverolles or Orpingtons for a happy flock.

Dual-purpose Breeds

Brahma

The majestic Brahma is an Asiatic breed believed to have originated in China. It was first introduced to England in the early 1800s, where it was considered so striking and desirable that it helped launch the infamous "Hen Fever" craze (see page 277). It has since contributed to the development of many breeds. Well into the twentieth century, breeders developed several color varieties of Brahma, and these now include the Light, Dark, Buff, Silver, Partridge, Blue, and Lavender. All color varieties feature heavily feathered legs and feet, and small pea combs, making them one of the most cold-hardy chicken breeds available. Brahmas are the "gentle giants" of the chicken world, even towering over other heavy breeds. Roosters reach an impressive 12 pounds (5.4 kilograms), and females may reach 9.5 pounds (4.3 kilograms) or more. Despite its size and beauty, this is a utilitarian dual-purpose breed. Brahma meat is very tender, though birds take nine months to reach full size. They are great egg layers, too, even through winter. Hens go broody often and are very good mothers. Brahmas are very gentle, quiet, and laidback, making them compatible with children, though their size can be intimidating. Roosters tend to be nonaggressive, but they're an imposing foe for anything they deem a threat. Brahmas thrive in smaller spaces, so are a fantastic choice for novice keepers and small backyards.

FACTFILE
Purpose: Dual purpose
Rarity: Uncommon; some varieties rare
Egg production: Good
Egg color: Brown
Origin: China
Size: Heavy
Temperament: Gentle, quiet

Above This Buff Brahma is a great specimen for showing off the heavily feathered legs and feet for which the breed is known.
Opposite The Light Brahma is primarily white with gray and black feathers. They look stunning but are also surprisingly hardy and gentle.

Buckeye

Named after the Buckeye State of Ohio, where it was developed, this beautiful and utilitarian dual-purpose breed was created in 1896 by Nettie Metcalf, who wanted a handsome bird that could withstand cold Ohio winters. The resulting Buckeye looks similar to the Rhode Island Red, but with darker, richer-red coloring and a small pea comb that resists frostbite well. Roosters may reach 9 pounds (4 kilograms), while hens can weigh up to 8 pounds (3.6 kilograms). Buckeyes are relatively uncommon, but they're an excellent dual-purpose breed, especially for those residing in colder climates. Hens are good layers of large brown eggs, and are likely to lay through winter. Buckeyes take confinement well, but they also thrive on open pasture, making them a good choice for homesteads and backyards alike.

FACTFILE
Purpose: Dual purpose
Rarity: Uncommon; some varieties rare
Egg production: Good
Egg color: Brown
Origin: United States
Size: Heavy
Temperament: Gentle, quiet

Above The Buckeye looks similar to a Rhode Island Red, but with even richer coloring and a more robust physique. **Opposite** Bred to survive cold Midwestern winters, the Buckeye's small pea comb is resistant to frostbite and they continue to thrive, even in snow and ice, as long as they have a well-insulated coop to retreat to for roosting (see page 226 for Heat and Cold Protection).

Sussex

The Sussex is among the most beautiful of the backyard chicken breeds, and is becoming very popular among keepers for its fancy looks and gentle demeanor. The breed comes in numerous color varieties, but the Speckled is by far the most popular, followed by Red, Light, Silver, Buff, and White. The Sussex is believed to have originated from ancient breeds—possibly dating back around 2,000 years—and was developed in the English county of the same name; legend states that each Sussex town developed its own unique color. By the late nineteenth century, the breed had become fully developed and standardized. It was so effective as a utilitarian breed that it was the primary commercial chicken in England until production hybrids took its place after World War II. The Sussex is a very heavy bird, with the roosters reaching up to 9 pounds (4 kilograms), and the hens up to 8 pounds (3.6 kilograms). It is the only chicken breed that comes in a Speckled color—a mahogany brown with lovely small white spots that are evenly spaced across its body, head, wings, and tail. The Sussex is cold-hardy, heat-tolerant, and takes confinement well, so it's an ideal breed for keepers looking for a friendly and pretty bird to add to their flock.

FACTFILE
Purpose: Dual purpose
Rarity: Common
Egg production: Very good
Egg color: Brown
Origin: England
Size: Heavy
Temperament: Calm, friendly

Above The Sussex is the only breed that comes in a Speckled variety, giving it an almost tweedy, mottled appearance. **Opposite** An exquisite bird in any color, the Sussex comes in Speckled (seen here), Red, Light, Silver, Buff, and White, with pink legs and peak and a bright-red comb.

Dual-purpose Breeds

Faverolles

The gentle Faverolles was developed in northern France in the 1860s as a homestead bird that could lay well through cold winters. Today, this breed maintains its excellent cold-hardiness and winter laying ability. It's also a lovely looking bird. The Salmon Faverolles is by far the most common variety, though all colors of this breed are considered rare. Salmon Faverolles hens are a lovely cream and salmon color, while roosters boast a cream neck and a dark red tail and wings. Both hens and roosters have short single combs, lightly feathered shanks and feet, and a fluffy beard-like muff under their chin. This is also one of the few chicken breeds that has five toes, rather than four. The Faverolles is moderately heavy, with roosters reaching up to 8 pounds (3.6 kilograms), and hens up to 6.5 pounds (2.9 kilograms). Despite its rarity, the Faverolles is earning a reputation for being an ideal breed for backyard chicken keepers, especially those who are interested in exhibition or breeding. It's an exceptionally quiet breed, and not prone to bullying or aggression. Because they are so docile, Faverolles should be kept with other gentle breeds.

FACTFILE
Purpose: Dual purpose
Rarity: Rare
Egg production: Good
Egg color: Cream
Origin: France
Size: Heavy
Temperament: Docile, gentle

Above The Salmon Faverolle (seen here) is the most common variety of this French breed. This is a gentle and quiet breed, known for being incredibly curious. **Opposite** The Faverolle is one of the few breeds that has five toes instead of four (as displayed here beneath the feathered shanks and feet).

Dual-purpose Breeds

Naked Neck

(Turken)

The Naked Neck—also known as the Turken—is an unusual-looking breed that lacks feathers on its neck, giving it a somewhat comical, turkey-like appearance. It also has about 50 percent fewer feathers overall than most breeds, due to a gene that results in fewer feather follicles. The Naked Neck was developed in Eastern Europe, and later perfected in Germany, to meet the desire for a hardy dual-purpose bird that was easier to pluck. Naked Necks come in several colors, including Red, White, Buff, and Black. They are a heavy breed, with roosters maxing out at about 8.5 pounds (3.9 kilograms), and hens 6.5 pounds (2.9 kilograms). Despite having so few feathers, Naked Necks are surprisingly tough and cold-hardy. They are also exceptionally well suited to warm climates. Naked Necks are intelligent birds, and very friendly toward humans. They free-range well, and are among the best layers of all dual-purpose breeds. Naked Necks are ideal for homesteaders, backyard keepers, and anyone who wants to add something unique to their flock.

FACTFILE
Purpose: Dual purpose
Rarity: Common
Egg production: Very good
Egg color: Brown
Origin: Eastern Europe, Germany
Size: Heavy
Temperament: Calm, friendly

Above In addition to the nickname of "Turken," the Naked Neck chicken is also called the Transylvanian chicken. Regardless of color, their trademarks are their featherless necks and single comb.
Opposite This Buff Naked Neck is a good example of their heavy, robust physique, making them tough and cold-hardy.

Dual-purpose Breeds

Jersey Giant

The imposing Jersey Giant is a slow-growing, very heavy breed, with roosters reaching up to an impressive 13 pounds (5.9 kilograms) or more, and hens reaching around 11 pounds (5 kilograms). This breed was developed in New Jersey by John and Thomas Black in the late nineteenth century as a roasting-meat bird to rival the turkey, and with good egg-laying ability. Jersey Giants come in three colors: the original Black, the later White, and the most recently developed Blue (the latter two are considered rare). All three varieties have feathers that sit tightly against their bodies, making them easy to care for, and tolerant of cold weather. Jersey Giants are stately birds, with a calm temperament. Hens are dependable layers of large, brown eggs, and tend to go broody. They make excellent mothers. The roosters are generally friendly, but they're also excellent defenders of their flock, thanks to their size. Hawks are believed to avoid Black Jersey Giants, so these large black birds are frequently kept in free-ranging flocks as a deterrent.

FACTFILE
Purpose: Dual purpose
Rarity: Common
Egg production: Good
Egg color: Brown
Origin: United States
Size: Heavy
Temperament: Friendly, docile

Above The most common color for this impressively sized bird is black (as displayed by this rooster). Jersey Giant roosters are generally friendly. **Opposite** With tightly packed feathers and an imposing build, Jersey Giants are easy to care for and survive cold climates. Their black feathers can display a gorgeous purple iridescence.

Dual-purpose Breeds

Wyandotte

Wyandottes are among the most popular of the dual-purpose breeds. Originally developed in the northeastern United States in the 1870s, and named in honor of the Wyandot people (or Huron) of the region, the breed is famous for its laying ability, docile temperament, and stunningly beautiful feathers. Their feather pattern is very similar to the ornamental Sebright bantam (see page 150), and in fact the Wyandotte was originally known as the American Sebright. Wyandottes also come in many color varieties, including Silver Laced, Gold Laced, Buff, Partridge, Colombian, and Blue Laced Red. These are stout, cold-hardy birds, with relatively short legs, dense feathers, and rose combs. Wyandottes are a heavy breed, too, with roosters reaching 8.5 pounds (3.9 kilograms), and hens reaching 7 pounds (3.2 kilograms). They're an excellent choice for backyard flocks, including families with children.

FACTFILE
Purpose: Dual purpose
Rarity: Common
Egg production: Good
Egg color: Brown
Origin: United States
Size: Heavy
Temperament: Friendly, docile

Above This Silver Laced Wyandotte displays the stout body, dense feathers, rose comb, and short legs typical of the breed. **Opposite** Wyandottes are a stunning breed in any color. Seen here in Gold Laced, Wyandottes are famous for their egg-laying ability, but they're equally good companions, including for families with children.
Following spread Two adult Silver Laced Wyandottes enjoy the free-range lifestyle.

Dual-purpose Breeds

Bielefelder

The Bielefelder, or Bielefelder Kennhuhn, is a newer dual-purpose breed. Developed in the 1970s in the Bielefeld area of northwest Germany, its reputation as a perfect backyard chicken is growing rapidly. The Bielefelder has a rare and very useful characteristic: it is an autosexing breed, meaning male chicks are easily distinguished from females on the basis of their color alone. Unlike "sex-linked" hybrids (see page 124), autosexed breeds retain this characteristic with every generation. Bielefelders are very pretty birds; the roosters are especially beautiful, with laced feathers in colors ranging from bright gold, to rust-red and silver. The hens are primarily rust-red with subtle silver lacing. This breed is heavier than most, with roosters reaching up to 12 pounds (5.4 kilograms), and hens reaching nearly 10 pounds (4.5 kilograms). Bielefelders are beginner-friendly, and well suited to families with children.

FACTFILE
Purpose: Dual purpose
Rarity: Uncommon
Egg production: Good
Egg color: Brown
Origin: Germany
Size: Heavy
Temperament: Friendly, docile

Above This Bielefelder rooster will have been distinguishable from the females since birth, based on color alone. A lot of people like the Bielefelder because of its auto-sexing ability, which means that the females and the males hatch in different color patterns. A male chick will be lighter and has a spot on his head that is yellow or white. A female chick will have a stripe on her back. **Opposite** The Bielefelder's reputation as the perfect backyard chicken has grown rapidly since this newer breed was developed in the 1970s.

Dominique

The Dominique is the oldest of all the American breeds, originating more than two hundred years ago in New England. It was a favorite among early settlers in the region, who took these chickens with them as they moved west. Today, the Dominique is considered a rare breed. Thanks to its dark gray and white barring, it is frequently confused with the more common Plymouth Barred Rock (see page 60)—its large red rose comb is what sets the Dominique apart. The Dominique is also a little smaller, with roosters reaching 7 pounds (3.2 kilograms), and hens 6.5 pounds (2.9 kilograms). Hens are good layers of large brown eggs, and unlike most other dual-purpose breeds, rarely go broody. Dominiques today retain the same qualities that made them so popular with early pioneering families—they are gentle and friendly, easy to feed and handle, and well suited to the backyard.

FACTFILE
Purpose: Dual purpose
Rarity: Rare
Egg production: Good
Egg color: Brown
Origin: United States
Size: Medium
Temperament: Friendly, docile

Above Like the Plymouth Barred Rock, the Dominique also has gray-and-white barring, but the large red rose comb is what makes it distinctive. **Opposite** In addition to being hardy and friendly, Dominiques are often considered ideal backyard companions and good with other breeds in a mixed flock.

Dual-purpose Breeds

Welsummer

Welsummers are a classic dual-purpose breed, developed after World War I near the village of Welsum in the Netherlands, from a variety of breeds thought to have included Rhode Island Reds (see page 59), Leghorns (see page 119), Barnevelders (see page 84) Cochins (see page 149), and Wyandottes (see page 74). Hens are very good layers of large brown eggs. Thanks to their somewhat smaller size, they also have a good feed-to-egg ratio, making them a fine choice for heritage egg farms. Welsummer hens are soft brown in color, with gray stippling on their backs and wings, and a lighter gold penciled neck. Roosters differ greatly in appearance from the hens, with rust-red hackle and saddle feathers draped over a black body. Their tail is also black, with an iridescent sheen. Welsummer roosters reach about 7 pounds (3.2 kilograms); hens reach around 6 pounds (2.7 kilograms).

FACTFILE

Purpose: Dual purpose
Rarity: Common
Egg production: Very good
Egg color: Brown
Origin: Netherlands
Size: Medium
Temperament: Quiet, docile

Above The Welsummer rooster's appearance is considerably different than the hens, more so than in most other breeds. The rooster's feathers are a mix of black, rust-red, and orange, while the hen's are a warm brown all over. **Opposite** The Welsummer breed (both roosters and hens) are smaller-than-average fullsize breeds. Their compact stature and egg-laying capability make them an excellent choice for smaller farms.

Barnevelder

This breed was developed in the area around the Dutch town of Barneveld. Its origins can be traced back to the Asiatic breeds introduced here in the mid-nineteenth century, which were then crossed with Dutch landrace chickens; the Barnevelder was developed as a good layer of dark-brown eggs, and received its present name in 1910. Its eggs are indeed among the darkest brown of any breed, perhaps only surpassed by the eggs laid by Copper Marans (see page 112). Barnevelders remain very popular in their native the Netherlands, and have become more familiar in recent years in Britain and the United States. Sometimes affectionately called "Barnies" by their fans, they are gaining a reputation for their curious, friendly personality. The Barnevelder is a very pretty bird, and comes in various colors and patterns, including Partridge, Double Laced Blue, and Double Laced Silver. Roosters are heavy, reaching up to 8 pounds (3.6 kilograms), while hens may reach 6.5 pounds (2.9 kilograms). They have yellow legs and beaks, red faces, and small, red single combs. Barnevelders are very adaptable and great birds for beginner chicken keepers.

FACTFILE
Purpose: Dual purpose
Rarity: Uncommon
Egg production: Very good
Egg color: Brown
Origin: Netherlands
Size: Heavy
Temperament: Quiet, docile

Above The Barnevelder breed is tremendously popular in the Netherlands (where it originated) and is gaining traction in the United States and Britain as an all-around great backyard breed. **Opposite** The Double Laced Barnevelder displays a particularly stunning set of wing feathers with exquisite patterning on each feather.

Dual-purpose Breeds

New Hampshire Red

The New Hampshire Red is not as famous as the Rhode Island Red (see page 59), but it's just as valuable as a dual-purpose breed, and an excellent choice for small farms and homesteads. It was developed in the early twentieth century in New Hampshire, from Rhode Island Red stock, with the aim of creating a faster-maturing version of that bird. The breeding efforts were successful, and the New Hampshire Red remains a popular choice among keepers today for its fast growth rate, good laying capability and especially good meat production. Like the Rhode Island Red, there is some variation within this breed, depending on the strain. Egg-producing strains tend to be a little smaller, with roosters reaching only 6 pounds (2.7 kilograms) or so, while the meat-producing strains can reach closer to 8 pounds (3.6 kilograms). All strains share the same yellow legs and beak, and mahogany red coloring. The New Hampshire Red has a reputation for being aggressive, so it's generally not recommended for mixed flocks or homes with small children.

FACTFILE
Purpose: Dual purpose
Rarity: Common
Egg production: Very good
Egg color: Brown
Origin: United States
Size: Medium
Temperament: Active, feisty

Above The New Hampshire Red is best known for its fast growth rate and good meat production. **Opposite** This breed is known for being active and feisty, and even aggressive, so it is best to take this into serious consideration. It is not recommended for mixed flocks or as a companionable bird, especially with small children.

Dorking

The Dorking is thought to be one of the oldest breeds of domesticated poultry, with origins dating back thousands of years. There are depictions of birds closely resembling the Dorking in ancient Roman art and literature, although it's uncertain whether the invading Romans first brought these food-producing chickens to the British Isles with them, or whether they encountered them when they arrived. In any case, the chicken takes its name from the Surrey town of Dorking, in southeast England, and much of its development took place in that country. The Dorking is considered a "foundation" breed, with important genes that have contributed to many more modern breeds. It's a unique-looking chicken, with short legs and a stout body. It also has five toes on each foot. Roosters weigh in at about 9 pounds (4 kilograms), and hens at 7 pounds (3.2 kilograms). Hens are dependable layers of cream-colored eggs and are unlikely to go broody. This breed comes in a wide variety of colors, including Red, White, Colored, Cuckoo, and Silver Gray, with the latter generally being easier to find. Dorkings may have a single comb or rose comb, depending on the variety. Their calm, independent nature makes them perfect for a small homestead or backyard flock.

FACTFILE
Purpose: Dual purpose
Rarity: Very rare
Egg production: Good
Egg color: Cream
Origin: England
Size: Heavy
Temperament: Quiet, independent

Above The Dorking is one of the oldest breeds of domesticated chicken, and this stunningly contrasted Silver Gray Dorking has graced many antiquated paintings in its time. Silver Gray Dorkings have a black breast and tail and white hackle and saddle. The Dorking was recognized by the American Poultry Association in 1874.
Opposite The Red Dorking is considerably different in color to the more common Silver Gray. This coloration was popular in the nineteenth century but dwindled in popularity. Fortunately, this color variety was kept alive by a small handful of breeders and is now on the rise. **Following spread** When choosing a mixed flock, it is essential to consider the temperament of each breed to ensure they can peacefully live together without aggression.

Dual-purpose Breeds

Langshan

The majestic Langshan is an ancient breed of chicken that originates from the area around Langshan (Wolf Hill), in China's Jiangsu province. It was first imported to Europe in the early nineteenth century, where it later contributed to the development of further breeds such as the German Langshan and the Orpington (see page 53). Langshans resemble standard-sized Cochins (see page 149), but are taller in stature and weigh a little less, with roosters reaching around 9.5 pounds (4.3 kilograms) and hens 7.5 pounds (3.4 kilograms). Their shanks and feet are feathered, though not to the degree of a Cochin. Langshans are only available in solid White, Black, and Blue. They have a "gentle giant" appearance and temperament, similar to the Brahma (see page 63). Langshans are surprisingly active and agile for their size, though, and they free-range well. They are a great choice for beginners and families, as well as those interested in rare breeds.

FACTFILE
Purpose: Dual purpose
Rarity: Rare
Egg production: Good
Egg color: Brown
Origin: China
Size: Heavy
Temperament: Docile, calm

Above The Langshan is a great mix of a rare, ancient, and stunning breed, while also being perfect for beginners and families. **Opposite** This Langshan rooster shows off his warm black plumage and bright-red comb. **Following spread** These Langhans are happily free-ranging in a mixed flock of other docile breeds.

Scots Dumpy

The adorably named Scots Dumpy is a rare heritage breed of chicken that originates from the Scottish Highlands; references to a chicken of its appearance date back centuries. The Scots Dumpy somewhat resembles the more common Plymouth Barred Rock (see page 60), but is distinguished by its extremely short legs—the result of a dwarfing gene—which mean that its belly nearly touches the ground. Despite its short stature, the Scots Dumpy is a heavy breed, with roosters capable of reaching 7 pounds (3.2 kilograms). Hens are good layers of cream-colored eggs. They tend to go broody often and make excellent mothers. While the barred Cuckoo color is most common, Scots Dumpys are also found in Black, White, Brown, and Silver. The Scots Dumpy is a tough, cold-hardy chicken, but some care must be taken to ensure they can move around easily, especially in mud and snow. Otherwise these friendly birds are an excellent choice for a backyard flock.

FACTFILE
Purpose: Dual purpose
Rarity: Very rare
Egg production: Good
Egg color: Cream
Origin: Scotland
Size: Heavy
Temperament: Friendly, docile

Above The trademark short legs of the breed are caused by a dwarfing gene. In spite of its belly being close to the ground, it is a surprisingly tough breed. (See page 226 for Heat and Cold Protection for more detail.) **Opposite** This Scots Dumpy rooster displays the beautiful coloration of the more common barred Cuckoo color, seen here in a farm in the Cotswolds region of England.

Minorca

The Minorca is a Mediterranean heritage breed; its ancestors came from the Balearic island of Menorca, but later development likely took place in England, which is where it eventually became a standard breed. Like their relatives on the Iberian Peninsula, Minorcas are known for their excellent heat tolerance, making them a great choice for keepers in very warm climates. The Minorca was also once famous for its excellent laying ability and large white eggs, but in recent decades it's been bred increasingly for exhibition, and its laying ability has diluted somewhat as a result. Still, this breed is a good layer and hens are unlikely to go broody. This is a handsome breed, with very large white earlobes that contrast sharply with red single combs and a sleek solid-colored body. The Minorca is usually found in Black, but Blue, Buff, and White birds have also been developed. In addition, a more cold-hardy rose-combed variety can be found, though it's less common. Minorca roosters can reach up to 9 pounds (4 kilograms), and hens up to 7.5 pounds (3.4 kilograms). They are generally friendly, but some are reported to have a feisty personality.

FACTFILE
Purpose: Dual purpose
Rarity: Uncommon
Egg production: Very good
Egg color: White
Origin: Spain
Size: Heavy
Temperament: Feisty, flighty

Above These stunning birds are now mostly bred for exhibition, but where kept in flocks, they are great for warmer climates (seen here as the hardier rose-combed variety). **Opposite** Minorcas are stunning birds, with large white earlobes and contrasting red combs. This Minorca cockerel is an exhibition bird, seen here at the Three Counties Showground near Malvern in Worcestershire, England.

FANCY EGG BREEDS

Fancy egg-layers—breeds and mixes famous for their beautifully colored eggs—have risen quickly in popularity in recent years. Going far beyond the traditional light brown or white, the eggshells of these breeds include bright blue, pink, sage green, and chocolate brown. Anyone interested in assembling their own share-worthy rainbow basket of eggs will want to acquire at least a couple of these breeds.

Shell color is a fascinating result of genetics and, to a lesser degree, environment. Eggs are made of calcium carbonate so they always start out white. The pigments that make up the great variety of final colors are then deposited on the egg as it moves through the oviduct. Chickens generally produce one of two pigments, depending on the breed. Protoporphyrin produces brown-tinted eggs, while oocyanin produces shades of blue. Because it is applied after the eggshell develops, the pigment color is richest on the outermost layers of the eggshell. White egg-layers do not produce any pigment at all. Many fancy egg-layers, such as the Easter Eggers (see page 103) and Olive Eggers (see page 107), may inherit both protoporphyrin and oocyanin, resulting in dark green eggs.

Opposite Araucanas (seen here) from Chile are an extremely rare breed that produces striking light-blue eggs. **Above** This trio of eggs are from the Easter Egger breed, so named for the range of colors produced.

Easter Egger / Americana

aster Eggers (also known as Americanas) are not technically a breed at all, but any hybrid resulting from crossing a colored egg-layer with a brown egg-layer. Because of this variation in makeup, there is no standard appearance—they may appear in many colors and patterns—although most examples of this designer chicken derive from the purebred Ameraucana (see page 108), which gives many of them a hawk-like face and bearded muff. As the name suggests, Easter Eggers are not standardized in their egg color, either; they may lay varying shades of white, pink, brown, green, or blue, although the color of egg that each hen lays remains the same throughout her life. Easter Egger roosters tend to reach around 5 pounds (2.3 kilograms), while hens are around 4 pounds (1.8 kilograms). They are generally very friendly birds, and easy to care for, so are especially popular among families with children.

FACTFILE
Purpose: Eggs
Rarity: Common
Egg production: Varies; usually very good
Egg color: Varies; any color possible
Origin: United States
Size: Medium
Temperament: Varies; usually active and friendly

Above Easter Eggers have no regular appearance, as they are not a standard breed. Instead, they are bred by crossing a colored egg-layer with a brown egg-layer, to produce a range of colored-egg laying hens.
Opposite Easter Eggers are common, generally small birds that are easy to look after. They are friendly yet active, so they can be a good member of a household with children.

Araucana

The Araucana is a fascinating breed that originated in Chile's Araucanía region, where it remained mostly isolated for centuries. Many believe that this breed is a "smoking gun" that points to Polynesians reaching the South American coast, where they introduced their own chickens to the continent decades prior to the arrival of the Spanish. Regardless of its origin, years of isolation resulted in a landrace chicken with several unique traits. In fact, when outsiders first stumbled upon it in the early twentieth century, they initially believed it to be a different species of poultry entirely. Over the next few decades, a formal breed was developed to highlight these unique traits: light-blue eggs, ear tufts, and a short, tailless rump. Unfortunately, some of these traits are also associated with embryo development and many Araucana chicks die before hatching. For this reason, the Araucana is a very rare breed, and only available through private breeders. Those that are lucky enough to meet a true Araucana will see a unique-looking, lightweight bird—roosters reach only 5 pounds (2.3 kilograms)—sporting fluffy ear tufts, an upright stance, shortened spine, and no tail. They come in at least twenty color varieties, including Blue, Black, White, and Silver.

FACTFILE
Purpose: Eggs, ornamental
Rarity: Very rare
Egg production: Good
Egg color: Blue
Origin: Chile
Size: Lightweight
Temperament: Independent, quiet

Above When non-Chileans first came across the Araucana, they did not believe it to be a chicken at all. Its ear tufts and tailless rump give it an unusual look in the chicken world, as seen here in the Silver Gray Araucana. **Opposite** While Araucanas are lovely, independent birds, the traits they've been bred for have meant they are delicate to hatch and now very rare.

Fancy Egg Breeds

Olive Egger

The Olive Egger is a mixed-breed chicken, similar to the Easter Egger (see page 103), but it was developed to lay only olive-colored eggs. These birds are generally a cross between a blue egg-layer and a brown egg-layer, bred over time to result in consistently green eggs—the darker green, the better!—although shades range from pale green through to dark olive, depending on the genes passed on by the brown-egg layer. Olive Eggers vary widely in appearance, depending on the breeds used to create them, and there is no standard. They may appear similar to Ameraucanas, but traits from other breeds may be apparent as well. Generally, the Olive Egger is a medium-sized bird, with roosters reaching between 6.5 to 8 pounds (2.9–3.6 kilograms). Egg-laying ability may also vary, but thanks to their parentage, they are usually very good egg layers. They are generally friendly birds, and are a popular choice among families with children, who will delight in gathering their beautifully colored eggs.

FACTFILE
Purpose: Eggs, ornamental
Rarity: Uncommon
Egg production: Varies; usually very good
Egg color: Varies; usually dark green
Origin: United States
Size: Medium
Temperament: Varies; usually active and friendly

Above This adorable Olive Egger chick can grow up to be of any color or body shape, depending on the breeds of its parents.
Opposite Like their cousin the Easter Egger, the Olive Egger has no standardized appearance. Instead, their common thread is the ability to lay consistently green eggs. They are friendly and can be a good choice for families with children.

Ameraucana

Not to be confused with the Americana, or Easter Egger (see page 103) the Ameraucana is a registered chicken breed, having been added to the American Poultry Association's Standard of Perfection in 1984. It was developed in the United States in the 1970s from Araucanas (see page 104), with the intention of creating a healthier counterpart to its Chilean ancestor, which is why its name combines "America" and "Araucana." It is an attractive breed that reliably lays light blue eggs and it comes in eight recognized colors: Black, Blue, Blue Wheaten, Wheaten, Brown Red, Buff, Silver, and White, along with newer colors like Lavender, Chocolate, and Splash. It always sports a fluffy muff under its beak. This is a medium-sized breed, with roosters reaching up to 7 pounds (3.2 kilograms), and hens up to 6.5 pounds (2.9 kilograms). For those living in small spaces, the Ameraucana is also found in bantam size, though these are harder to find. Unlike the Easter Egger and Olive Egger, which may be similar in appearance, the Ameraucana always lays light blue eggs only.

FACTFILE
Purpose: Eggs
Rarity: Common; some colors rare
Egg production: Very good
Egg color: Blue
Origin: United States
Size: Medium
Temperament: Friendly, active

Above The medium-sized breed of the Ameraucana comes in a wide range of colors, seen here in the recognized coloration of Brown Red. They can also be found in bantam size. **Opposite** The Ameraucana displays the same charming fluffy beard of the Araucana but was bred as a healthier variation.

Fancy Egg Breeds

Legbar

The Legbar is a fantastic breed for keepers wishing to have a bounty of fresh, bright blue eggs every day. This breed is an excellent layer, on a par with the more popular Leghorn (see page 119). What sets the Legbar apart, though, aside from its eggs, is its rare autosexing attribute, which makes males and females easily identifiable at hatching. Developed in the early twentieth century, the Legbar was the second autosexing chicken created by Reginald Punnett and Michael Pease at the Genetical Institute of Cambridge University—in this case by cross-breeding Barred Plymouth Rocks (see page 60) with Leghorns, to develop a production breed capable of sexing at hatch. While initially popular, blue eggs later fell out of favor and commercial hybrids quickly rose in popularity, bringing the Legbar close to extinction. However, it's now one of the most sought-after backyard chicken breeds, and numbers are recovering quickly. Legbars are active and alert birds with an almost comical appearance, thanks to the crest of feathers and floppy red combs on their heads. They are a little larger than the Leghorn, with roosters reaching up to 7 pounds (3.2 kilograms). They come in Gold, Silver, and, most commonly, Cream. Legbars are exceptional foragers and thrive in free-range environments, but they do well in backyards, too.

FACTFILE
Purpose: Eggs
Rarity: Rare
Egg production: Excellent
Egg color: Blue
Origin: England
Size: Medium
Temperament: Friendly, active

Above The Cream Legbar is the most common coloration, which includes a range of gray, black, and even rust feathers. This Gold Legbar hen displays a salmon breast. **Opposite** In addition to being beautiful birds, Legbars are excellent layers of blue eggs.

Fancy Egg Breeds

Marans

The Marans is an old French breed that was originally developed to be a dual-purpose chicken. Keepers today, however, covet the Marans for its beautiful, dark chocolate-colored eggs. The Marans comes in many colors, with the Silver Cuckoo, Gold Cuckoo, White, and Copper Black varieties making up just four of the ten colors recognized by the French breed standard, for example. The French Copper Black Marans lays the darkest-brown eggs of any chicken, and is currently in high demand from keepers looking to complete the perfect rainbow egg basket. Depending on variety and location, Marans may or may not be feather-footed. This is due to their development. These chickens were bred for their superior dark eggs; appearance was treated more as an afterthought, and not formalized until much later. By that time, several strains of the Marans were being bred in different regions. In general, the French Marans breeds are recognized as feather-footed, while the English Marans are not. All varieties are compact and densely feathered, with large, red, single combs that can be prone to frostbite in cold climates. Aside from their dark brown eggs, all Marans are lovely backyard birds that get along well with children and other chickens. They are friendly, and relatively quiet, too, making them an ideal choice when neighbors are in close proximity.

FACTFILE
Purpose: Dual purpose
Rarity: Uncommon; some varieties rare
Egg production: Good
Egg color: Dark brown
Origin: France
Size: Heavy
Temperament: Friendly, curious

Above The Black Marans (seen here as a rooster) has shiny black plumage with a green sheen, with red comb and ear lobes.
Opposite This Copper Black Marans is known for laying the darkest-brown eggs of any known chicken. As long as you don't live in a cold climate, Marans could be a great choice for a mixed flock, families with children, and those with nearer neighbors.

Whiting True Blue & Green

The Whiting True Blue and Whiting True Green are relatively new hybrid breeds. They are named for the breeds' developer, renowned geneticist Dr. Tom Whiting, whose initial aim was to create chickens that could provide feathers for use in fly fishing. The two variants lay plenty of bright blue and light green eggs respectively. Unlike Easter Eggers or Olive Eggers (see pages 103 and 107), the Whiting True Blue and True Green are a fully developed breed that always lay the same-colored egg. Because this breed was developed for egg color only, the birds themselves can vary in appearance. Some individuals may have muffs like the Ameraucana (see page 108); others may not. Whitings also come in a wide variety of colors. The Whiting True Greens tend to be more uniform in appearance, with no muff and colors ranging from rust red to a lighter brown and white. All Whitings are medium-sized birds, with roosters reaching up to 7 pounds (3.2 kilograms), and hens reaching up to 4 pounds (1.8 kilograms). They are friendly, active birds that forage well and do well in homes with small children.

FACTFILE
Purpose: Eggs
Rarity: Uncommon
Egg production: Excellent
Egg color: Blue or green
Origin: United States
Size: Medium
Temperament: Friendly, docile

Above Whitings do not have a standardized appearance, as they are bred for egg color. Some individuals may have faces more like the Ameraucana (such as this individual). **Opposite** Whiting True Greens often are rust red to lighter brown and without a muff, such as this. Their eggs of these hybrid breeds are always beautiful shades of blue or green.

PRODUCTION LAYERS

Production-layer breeds and hybrids were developed for one thing only: to produce as many eggs as possible for the least amount of feed and time. These are the commercial layers that are raised by the millions on large poultry farms worldwide. And of all the breeds and hybrids listed in this section, only the Leghorn existed before the mid-twentieth century. Production-layer breeds are not just meant for large-scale egg farms; they are also very popular birds among backyard chicken keepers and homesteaders because they mature quickly, lay prolifically, and are well suited to urban backyards and other small spaces.

On farms where egg production is a focus, production layers are raised for two to four years, when egg production is at its peak. After their second or third laying season, egg production tends to drop dramatically with these breeds, so once production begins to decline, they are either retired, rehomed, or in some cases butchered, and a new generation of production layers is raised. For chicken keepers and homesteaders looking to generate income and provide their community with locally sourced, nutritious eggs, these production layers are an excellent choice.

Production-layer breeds and hybrids have varied temperaments, ranging from extremely friendly and docile, to nervous and flighty. Hens rarely go broody, and generally don't make good mothers. Roosters are known for being aggressive, though there are exceptions. They are good choices for backyard flocks, but they can be more prone to reproductive health issues and shortened lifespans, particularly if they've been rescued from a factory farm.

Opposite Production-layer breeds are also well suited for backyard chicken keepers wishing for a reliable source of eggs. **Above** Leghorns have been considered the pinnacle of production layers for generations.

Production Layers

Leghorn

Originating in Italy, the Leghorn has been the gold standard for egg laying for generations. It was first introduced to the United States in the early 1800s, and has been ranked among the best layers since. It is a lightweight, medium-bodied bird, with roosters and hens reaching only 5 pounds at most. The Leghorn is available in several strains and varieties, including the high-producing White, which is still used on factory farms today for its abundance of large, white eggs. Leghorns are available in several heritage and ornamental varieties as well, such as the Brown, Red, and Silver. Most varieties sport a large single comb, but showy rose combs are also available from hatcheries and breeders. All varieties of Leghorn have yellow legs and beak, and white earlobes. They also always lay pure white eggs. Like other Mediterranean breeds, the Leghorn is exceptionally heat tolerant, but prone to frostbite and not very cold hardy. Backyard chicken keepers and farmers tend to be drawn to the heritage lines of Leghorn, which are just as healthy and long-lived as other heritage breeds. Despite their use in factory farms, Leghorns are active and intelligent birds, and thrive as foragers on free-range farms. Leghorns are famous for being nervous and flighty, and may not be the best choice for beginner keepers, or those who are looking for a friendly chicken.

FACTFILE
Purpose: Eggs
Rarity: Very common, some varieties rare
Egg production: Excellent
Egg color: White
Origin: Italy
Size: Lightweight
Temperament: Active, flighty

Above This young White Leghorn displays the standard yellow legs and beak, red comb, and white ear lobes. **Opposite** For backyard keepers looking for prolific egg-layers who thrive by foraging on free-range farms, this is your bird. But these birds can be nervous and flighty, so not ideal for those looking for a companionable family pet.

Production Red

The Production Red is a hybrid that was developed from the Rhode Island Red (see page 59) for abundant egg production. The specific breeds involved vary, but generally include Rhode Island Whites, Plymouth Rocks, or New Hampshire Reds (see page 87). Unlike many other production hybrids, Production Reds are not sex linked; females and males look alike at hatching. Production Reds are very similar to the Rhode Island Red, with a yellow beak and legs, and bright red combs. Their coloring ranges from dark mahogany red to cream. This hybrid tends to lay better than purebred dual-purpose breeds, though it can reach a similar size. Production Reds are favored by homesteaders and small-scale farms for their high production of eggs and good meat-production capability. They're also good foragers and do well on open pasture. Production Reds have a tendency toward aggression, and the roosters in particular can be relentlessly territorial. They're therefore not recommended for beginners or backyard keepers with children.

FACTFILE
Purpose: Eggs
Rarity: Common
Egg production: Excellent
Egg color: Brown
Origin: United States
Size: Medium
Temperament: Territorial

Above Female and male Production Reds look alike at hatching and until they start to mature. **Opposite** With a very similar look to the Rhode Island Red, this breed can range in color from mahogany to cream, with a yellow beak and legs and red comb. **Following spread** Production Reds are a great choice for those looking for a steady supply of eggs and with open space for them to roam. However, they can be aggressive and territorial and are not suited to beginners or households with children.

SEX LINKS

Sex links are production-layer hybrids that are generally a mix between various heritage breeds to produce as many eggs as possible. Parent breeds tend to involve the Leghorn, Barred Rock, and Rhode Island Red. They get their name from their sex-specific characteristics that make males easily distinguishable from females at hatching—a very helpful attribute for egg-production farms. Unlike the rare "autosexing" characteristic, sex links do not pass down this trait to subsequent generations. The "breeds" of each sex-link hybrid are essentially brand names given to each by the hatcheries that produce them. They are not registered in any way, nor will they "breed true" should backyard keepers decide to breed them.

Opposite Like other sex-link hybrids, these Red Stars are easily distinguishable between the sexes at hatching. **Above** Red Star sex links are known for their cheerful disposition and outstanding egg production.

Amberlink

(Amber Star)

The Amberlink—also known as the Amber Star—is a type of Red sex-link hybrid (see page 124), usually involving the cross of Rhode Island Red or New Hampshire Red hens with White Plymouth Rock roosters (see pages 59, 87, and 60). The result tends to produce cream-bodied hens and roosters, with the latter sporting dark salmon markings. They are a popular hybrid among backyard chicken keepers, and are known for being docile, quiet, and friendly. Amberlinks are excellent layers of large brown eggs, even through winter.

FACTFILE
Purpose: Eggs
Rarity: Common
Egg production: Excellent
Egg color: Brown
Origin: United States
Size: Medium
Temperament: Docile, friendly

Above Amberlinks, or Amber Stars, have beautiful cream-white bodies. **Opposite** These hybrids are friendly and excellent egg-producers, making them a good choice for backyard chicken keepers seeking a regular supply of eggs from chickens who are not aggressive or flighty.

Red Star

*(Golden Comet, Gold Star, Buff Star,
Cinnamon Queen)*

Arguably the best known and most beloved backyard egg-layer, the Red Star is a type of Red sex link (see page 124) that goes by many names, depending on the hatchery that produces it and the breeds used in its development. These birds may be the cross of a White Plymouth Rock hen with a New Hampshire Red rooster, a Rhode Island White hen with a New Hampshire Red rooster, a Silver Laced Wyandotte hen with a Rhode Island Rooster, or a similar combination, as long as the hen is white with a silver gene, bred to a red-gene rooster. Like other sex links, they do not breed true. Even their sex-link attribute can be difficult to produce, and is highly dependent on the strain of parentage used. For this reason, true Red Star sex links like Golden Comets are rarely available outside of commercial hatcheries. Red Stars are easy to distinguish right at hatching. The males are nearly all white with reddish flecking; the females are a rusty red at hatch, maturing to a darker red with white wing and tail patches. Red Stars mature quickly, and hens may start to lay as early as twenty weeks. Red Star sizes vary, but they're generally large enough for meat production. No matter what name they go by, Red Stars are popular for a reason: the hens are known for their cheerful, docile personalities, and large brown eggs, making them great backyard chickens.

FACTFILE
Purpose: Eggs
Rarity: Common
Egg production: Excellent
Egg color: Brown
Origin: United States
Size: Medium
Temperament: Docile, friendly

Above Females hatch rusty red and mature quickly into a darker red with white wing and tail patches. **Opposite** While best known for their egg production, Red Stars are just as beloved for their cheerful, friendly demeanor. Many keepers love them for their personalities and production all in one.

Black Star

Another very popular backyard egg-layer, the Black Star is a type of Black sex-link hybrid (see page 124). Black Stars are generally produced by breeding a Barred Rock hen (see page 60) with a non-barred rooster—usually a Rhode Island Red or New Hampshire Red (see pages 59 and 87). Like other sex links, Black Stars don't breed true, so the offspring of two Black Stars will not technically be Black Stars themselves. For this reason, true Black Stars are rarely found outside of commercial hatcheries, though some private breeders may develop their own Black sex-link hybrids. Black Stars are easy to distinguish right at hatching because the "barred" gene, which presents itself as a light barring or spot on the head is only present on the male chromosome. Black Star hens have entirely black bodies with a yellow–gray bill and legs, a red face and comb, and light gold hackle feathers on their neck. Roosters are similar, but with much more gold and white on their necks and wings. Black Stars mature quickly, and hens may start to lay as early as twenty weeks. Sizes vary, but Black Stars are generally large enough for meat production. They're also nearly as popular as Red Stars (see page 129), thanks to their large brown eggs and docile personalities.

FACTFILE
Purpose: Eggs
Rarity: Common
Egg production: Excellent
Egg color: Brown
Origin: United States
Size: Medium
Temperament: Docile, friendly

Above Black Star hens are entirely black, with only tan-gold neck and breast feathers. **Opposite** Black Stars are produced by breeding a barred hen with a non-barred rooster. This rooster shows a beautiful gold and black barred pattern and is here clearly demonstrating his friendly personality.

ISA Brown

The ISA Brown is a type of Red Star hybrid (see page 129). Unlike most other Red Stars, the ISA Brown has had many different breeds and crosses go into its development. This is more of a dual-purpose hybrid breed, which is very popular on small farms and homesteads for its excellent egg and meat production. However, it's known for being territorial and aggressive, so it's generally not recommended for beginner keepers or families with children. The ISA Brown is a trademarked name and available only from Hubbard licensed hatcheries.

FACTFILE
Purpose: Eggs
Rarity: Common
Egg production: Excellent
Egg color: Brown
Origin: United States
Size: Medium
Temperament: Territorial

Above ISA Browns are a robust hybrid with excellent egg production. **Opposite** ISA Browns are recognized as excellent egg-layers and adaptable to different climates and arrangements. However, ISA Browns can be territorial so it is not advisable for novice keepers or children.

Production Layers: Sex Links

Sapphire Gem

The Sapphire Gem is a newer sex-link hybrid (see page 135) that's said to be developed from the Blue Plymouth Rock, Plymouth Barred Rock, and possibly other breeds. This production hybrid is rapidly growing in popularity due to its lovely Blue or Lavender plumage and large brown eggs. Sapphire Gem male chicks are easily distinguished at hatch from the females, thanks to a white dot on their heads, and they grow to be significantly lighter in color. Sapphire Gem hens, which are a lovely smoky gray, mature quickly, and may produce eggs as early as twenty-two weeks. They are medium-sized birds with yellow legs, dark red faces, and a single comb. Roosters reach about 7 pounds (3.2 kilograms) and hens around 6 pounds (2.7 kilograms). They tend to be very friendly and do well in mixed flocks. Like the ISA Brown (see page 133), the Sapphire Gem is a trademarked name so is only available through licensed hatcheries.

FACTFILE
Purpose: Eggs
Rarity: Uncommon
Egg production: Excellent
Egg color: Brown
Origin: Czech Republic
Size: Medium
Temperament: Docile, friendly

Above Originating in the Czech Republic, this newer hybrid is friendly and with excellent egg-laying capabilities. **Opposite** The pretty Lavender plumage (pale to warm gray) makes these birds stand out in a mixed flock, where they do well.

MEAT BIRDS (BROILERS)

Chicken breeds developed for meat production differ significantly from your typical backyard chicken, both in appearance and in development. Meat chicken breeds have been developed for the sole purpose of producing as much meat as possible in the shortest amount of time. The result is a stout, thick-bodied bird capable of growing many times faster than its traditional counterparts.

Commercial farms and meat-producing homesteads raise their meat birds (often called "broilers") for six to sixteen weeks, sometimes more, depending on the breed. Chicks reach butchering size as early as six weeks and are processed soon after. Smaller-scale farms and homesteads often opt for the healthier and hardier meat hybrids, preferring to raise them on pasture or covered tractors so the chicks get plenty of exercise and develop good muscle. This is also considered a more sustainable and humane way of raising meat birds. Because they grow so heavy and fast, meat-production chickens tend to suffer from a wide range of health issues, including cardiovascular problems, organ failure, and broken legs. Their quality of life and lifespan can be improved somewhat, however, through special care and feeding.

Meat-production breeds tend to be very docile and friendly, but are not recommended for backyard keepers due to their special needs and extensive health issues.

Opposite and Above These Red Rangers (above) and Broiler Hybrids (opposite) are meat-production birds, also known as broilers. While they can be friendly and do well on small farms, they are not recommended for novice keepers.

Meat Birds

Cornish Cross

The Cornish Cross is the best meat-producing bird in the world. It was developed in 1948 by crossing a Cornish with a White Plymouth Rock (see page 60) for this precise purpose, and the birds grow at astounding rates: roosters are capable of reaching 7 pounds (3.2 kilograms) in as little as six weeks, while females can reach the same size just a couple of weeks later. Due to their rapid growth rate, however, these chickens suffer the most significant health issues. Their cardiovascular systems in particular can be weakened, making them difficult to raise at high elevation or on pasture. And because they have a lot of trouble walking, Cornish Crosses are not good foragers so are best kept in safe confinement. Keepers wishing to keep Cornish Crosses alive and well beyond eight weeks must place them on a restricted diet and monitor their weight carefully. Cornish Cross chicks are often sold at feed stores alongside more common backyard breeds, so new keepers must take care not to take Cornish Cross chicks home by accident if they don't want to raise them for meat— these birds are only recommended for meat production, and are not well suited for typical backyard flocks.

FACTFILE
Purpose: Meat
Rarity: Very common
Egg production: Poor
Egg color: White
Origin: United States
Size: Medium
Temperament: Docile

Above The Cornish Cross can suffer health concerns and should not be kept as outdoor pets or looked after by keepers seeking anything other than a meat-production bird. **Opposite** This breed grows extremely quickly and was produced specifically for meat production.

Meat Birds

Broiler Hybrids

(Red Ranger, Heritage Broiler, Freedom Ranger)

Chicken keepers wishing to raise healthier and more active meat birds may prefer any number of broiler hybrids. (Most broiler varieties found in the US were developed there, and this tends to be the case for other countries, too.) These meat birds also grow extremely quickly, but still slower than the Cornish Cross (see page 139), with males reaching full butchering size by about fourteen weeks. They are known by different names, depending on the hatchery that developed them, but they generally share similar parentage—usually a Cornish Cross mixed with a dual-purpose heritage breed such as the New Hampshire Red (see page 87). Appearance may vary, but broiler hybrids tend to be a rust red in color, with red faces and combs, large breasts, and thick yellow legs to support their weight. They usually have a docile disposition. These broiler hybrids tend to do well on homesteads and small farms, and are also fit and agile enough to free-range on pasture. Broiler hybrids do experience health issues, though, if they're allowed to grow too long past their expected processing age, so they're not considered good backyard pets.

FACTFILE
Purpose: Meat
Rarity: Common
Egg production: Poor
Egg color: Varies; usually brown or white
Origin: United States
Size: Medium
Temperament: Docile, curious

Above While Broiler Hybrids (young pullet seen here) do grow quickly, it is not as quick as their cousin the Cornish Cross, so they are not as delicate. **Opposite** The appearance of Broiler Hybrids varies considerably, as seen with this full-grown white broiler.

BANTAMS & ORNAMENTAL BREEDS

When most people think of chickens—including backyard chickens—they tend to picture poultry primarily raised for meat and eggs, with any companionship offered being an added bonus. For much of the chicken's history, however, that wasn't the case at all. In fact, the idea of keeping chickens as companion animals is a concept that dates back thousands of years. The breeds listed in the following section were developed for reasons other than producing food, and some of them have origins even more ancient than the heritage breeds. Many were bred to be pets, or fancy ornamental birds to grace large estates and palace grounds. Others descend from ancient cockfighting breeds, and are treasured today for their beauty and compatability with humans. For keepers looking for flashy, fancy conversion-starters, these breeds are definitely worth considering.

Nearly all of the breeds listed here are classified as "bantams," meaning they are much smaller than your standard livestock chicken. They make wonderful pets, take up far less space than standard breeds, and don't require nearly as much feed. Most bantams are often scaled-down versions of standard-sized breeds, but some are only available in the bantam form. These are known as "true bantams." Bantam chickens' diminutive size makes them easy for children to handle, too, which is useful for youngsters who are interested in showing poultry through a local club. Some bantams' laying ability is also quite impressive, though their eggs are less than half the size of a standard chicken egg. Bantams generally weigh between 16 and 64 ounces (500 grams to 1.8 kilograms).

Above and opposite From the robust yet ornate Belgian d'Uccle Mille Fleur bantam to outrageously adorable Silkies, bantams and ornamental breeds are as diverse as the keepers themselves.

Old English Game Bantam

Most chicken keepers interested in exhibition at poultry shows will consider the Old English Game Bantam (OEGB)—possibly the most popular show chicken in the world! This small chicken is descended from the birds once used in England for cockfighting, and indeed poultry club standards still require roosters to go through a process called "dubbing," which removes the top of their comb as a nod to their fighting heritage. At least seventy colors have been developed so far—including Spangle, Lemon Blue, Red Pyle, and Silver Duckwing—along with other variations in appearance, depending on location and strain. All varieties are lively, curious birds with a friendly disposition. They are easy to handle, and make a lovely addition to backyard flocks. OEGBs are also exceptionally long-lived and healthy, and often used in the development of other breeds to add vitality. Hens are decent layers of small, cream-colored eggs. They tend to go broody, and make good mothers. The OEGB is a lightweight bird with a big personality. The roosters reach 2 pounds (90 grams) at maturity, and hens 1.5 pounds (70 grams). The Old English Game is also available in a larger standard size, with roosters weighing in at 4.5 pounds (2 kilograms), and hens reaching 4 pounds (1.8 kilograms), although the bantam is much more popular in the show ring. Roosters are proud and handsome, attentive to their hens, but prone to aggression, especially towards other roosters. Both hens and roosters are generally friendly toward humans, however, and make good pets.

FACTFILE
Purpose: Companionship, ornamental
Rarity: Common
Egg production: Good
Egg color: Cream
Origin: England
Size: Bantam and lightweight standard
Temperament: Curious, friendly

Above An incredible seventy-plus colors of Old English Game Bantams have been developed so far, and this Black OEGB is a beautiful example of the variety. **Opposite** All varieties of Old English Game Bantams are curious and friendly, making them easy to handle. They can be make good outdoor pets.

Modern Game Bantam

The Modern Game Bantam was developed from the Old English Game Bantam (OEGB; see page 145) as a fancy exhibition bird. This breed comes in a number of colors and varieties, but all share the same exaggerated upright posture and disproportionately long legs and necks, giving them a proud, swaggering look that makes them very popular at poultry shows. The Modern Game Bantam is a small, slender bird, with roosters reaching just 21 ounces (595 grams), and hens 20 ounces (570 grams). The Modern Game is also available in a larger standard size, with roosters weighing in at 6 pounds (2.7 kilograms), and hens at 4.5 pounds (2 kilograms). As with the OEGB, the bantam type is much more popular in the show ring than the standard type, and roosters are required to be dubbed before showing. Modern Games tolerate heat very well, but are susceptible to cold, and require supplemental heat in temperate climates. The breed is easy to handle and very friendly, and especially popular with children and keepers wanting an affectionate pet chicken.

FACTFILE
Purpose: Ornamental
Rarity: Common
Egg production: Poor
Egg color: White
Origin: England
Size: Bantam and lightweight standard
Temperament: Curious, friendly

Above Modern Game Bantams are both fancy exhibition birds and affectionate pets, ideal for keepers looking for a friendly yet unusual bird. **Opposite** This pair of White Modern Game Bantams are seen here at the 2014 Northeastern Poultry Congress, in West Springfield, Massachusetts.

Cochin

This very old breed originated in China. It first
appeared in America, by way of England, in the
mid-nineteenth century, and was initially known as
the Shangae (Shanghai) Fowl or Cochin China Fowl.
The bird's impressive size and abundant feathers made
it instantly popular with chicken fanciers, triggering
a craze known as "Hen Fever" on both sides of the
Atlantic (see page 277). Cochins are very fluffy birds,
with rounded rumps and heavily feathered legs and
feet, making them look a lot larger than they really
are. Still, the standard Cochins are large birds, with
roosters reaching up to 11 pounds (5 kilograms), and
hens reaching 8 pounds (3.6 kilograms). Cochin
bantams, which are becoming increasingly popular,
are significantly smaller, with roosters reaching only
32 ounces (900 grams), and hens 29 ounces (800
grams). Both standard and bantam Cochins come in
a wide variety of colors, ranging from Black, Barred,
and Blue, to Golden Laced, Mottled, and Partridge.
They are fair layers of medium-sized white eggs, but
they are primarily kept as companion and exhibition
birds. Cochin hens go broody easily and are
exceptional mothers, frequently adopting chicks from
other hens. Roosters are handsome birds that protect
their hens well, and are less likely to show aggression
toward humans. Cochin bantams are very quiet, and
ideally suited to small urban spaces, so they have
become a favorite with backyard chicken keepers.
They are also good with children, thanks to their
calm, docile personalities, so make wonderful pets.

FACTFILE
Purpose: Companionship, ornamental
Rarity: Common
Egg production: Fair
Egg color: White
Origin: China
Size: Heavy standard and bantam
Temperament: Curious, friendly

Above The Cochins given to Queen Victoria as a wedding gift were
believed to have started the "Hen Fever" that gave way to our
continued love for chickens to this day. **Opposite** This golden-hued
Buff Cochin Bantam is a beautiful example of the breed known for
being well suited to urban spaces because of their friendly, curious,
and quiet nature.

Bantams & Ornamental

Sebright

Ith roosters weighing only around 22 ounces (620 grams), and hens around 20 ounces (570 grams), Sebrights are a true bantam breed. They were developed as an ornamental breed by John Sebright in the early 1800s, and they have a striking appearance. Sebright roosters display a rare trait called "hen feathering," meaning they lack the fancy hackle, saddle, and tail feathers that distinguish most roosters from hens. This is the result of a rare gene, and it is found in only a couple of chicken breeds. Sebrights come in two colors—Golden and Silver—and both colors sport contrasting black lacing. They have a gray beak and legs, a red rose comb, and large, dark eyes. Hens are dependable layers of small white eggs, and rarely go broody. Sebrights are active and cheery birds, but they can be flighty. They are a must-have for keepers interested in exhibition, or those who wish to have a fancy backyard chicken flock.

FACTFILE
Rarity: Uncommon
Egg production: Poor
Egg color: White
Origin: England
Size: Bantam
Temperament: Curious, active

Above Seabrights come in just two colors, Golden (seen above) and Silver (seen opposite). Both show the striking lacing pattern on their feathers for which this breed has become known. **Opposite** These birds are known for having cheerful demeanors but they can be active and flighty, so bear this in mind when considering your preferred breeds for the environment.

Silkie

Of all the companion chicken breeds, the Silkie is hands-down the most popular. These are gentle, affectionate birds with the most adorable appearance. They are an ancient breed, having most likely originated in China, and depictions of them date back as far as the thirteenth century. Their most unique attribute is their plumage, which has a silky, hair–like texture (hence the name). Their exceptionally soft feathers cover them from the top of their heads right down to their feathered feet. Silkies also have black skin, small walnut combs, and five toes on each foot. They come in several recognized colors, including Black, White, Blue, Buff, Splash, and Partridge. They may also have a muff. Hens and roosters are so similar in appearance that they're difficult to sex until maturity. In the United States, Silkies are considered to be true bantams, regardless of size, whereas other regions make a distinction between large fowl and bantams. In the UK, for example, a large fowl rooster will reach around 4 pounds (1.8 kilograms) and a hen reaches around 3 pounds (1.4 kilograms); bantams are much smaller, averaging around 21 ounces (600 grams) and 18 ounces (500 grams) for roosters and hens respectively. Silkies require some special care. They should be protected from rain and other precipitation to protect their feathers, which are not very waterproof, and they tend to get chilled more easily than other breeds. Their large fluffy crests may also inhibit their vision, and they are not particularly good fliers. For small backyard keepers, however, the Silkie makes the perfect pet chicken.

FACTFILE

Purpose: Companionship, ornamental
Rarity: Common
Egg production: Fair
Egg color: White
Origin: China
Size: Bantam (sometimes also large fowl)
Temperament: Docile, affectionate

Above When most people hear "Bantam," the Silkie is what they picture: diminutive, fuzzy, and adorable. And the Silkie is all of these things and more. This adorable specimen is a young Partridge Silkie
Opposite This six-month-old White Silkie displays the fluffy crest, small size, feathered feet, and five toes iconic of the breed. However, these traits are not without consideration, as Silkies are neither waterproof nor good fliers, so they require extra care. But if the keeper can offer this, Silkies make for excellent pets. **Following spread** Silkies come in a a wide range of colors, all soft. The White and Black Silkies seen here are demonstrating their affectionate side.

Sultan

This ornamental breed originated in Turkey, where it was favored by royalty (its Turkish name, *serai taook*, means "Sultan's fowl"). It was exported to England in the mid-nineteenth century, and reached the United States soon after. Sultans are somewhat similar in appearance to Silkies (see page 153), with five-toed feet, feathered legs, and a large feathered crest. Their feathers, however, lack the hair-like quality that distinguishes a Silkie. Sultans are a pure, snowy white, with a pink beak and a bright red V-shaped comb. Some Blue and Black varieties are available, too, but these are exceptionally rare. They are close in size to the Polish (see page 166), with roosters reaching up to 6 pounds (2.7 kilograms), and hens up to 4 pounds (1.8 kilograms). Although the Livestock Conservancy lists the Sultan's status as critical, rising demand for the breed as a backyard chicken means that numbers are improving. It has a friendly disposition and, like the Silkie, makes a wonderful backyard pet.

FACTFILE

Purpose: Companionship, ornamental
Rarity: Very rare
Egg production: Fair
Egg color: White
Origin: Turkey
Size: Medium
Temperament: Docile, affectionate

Above Sultans have a distinctive feathered crest, feathered legs, and five toes. **Opposite** The Sultan was originally favored by royalty, and this beautiful individual is certainly demonstrating a regal air. This breed is friendly and affectionate.

Belgian d'Uccle

The adorable Belgian d'Uccle, or Barbu ("Bearded") d'Uccle, is a true bantam breed, characterized by its unique feather patterns and heavily feathered feet. Although it was formally developed in Belgium by Michel Van Gelder in the late 1800s, this breed is thought to have originated in Japan. The d'Uccle is highly prized as a backyard companion. It is a true bantam, weighing around 26 ounces (740 grams) for roosters and 22 ounces (625 grams) for hens, with a friendly, outgoing personality. This breed is widely available in two colors—Mille Fleur and Porcelain—and nearly 20 additional colors have been developed to date, including Black Mottled, Blue, Buff, and White. The have red single combs, wide upright tails, and a fluffy muff under the beak. Belgian d'Uccles are good layers of small, white eggs. They are also an excellent choice for backyard chicken keepers who are looking for a beautiful and friendly bird. They are easy keepers, love to be held, and are very good with children and beginners.

FACTFILE
Purpose: Companionship, ornamental
Rarity: Uncommon
Egg production: Good
Egg color: Cream
Origin: Belgium
Size: Bantam
Temperament: Docile, affectionate

Above The Belgian d'Uccle is a true bantam. The Black Mottled coloration (seen here) is less common. The Mille Fleur can be seen on page 143. The Porcelain coloration (opposite) appears almost silver.
Opposite This breed is not only adorable and affectionate, they love to be held, making them ideal for keepers looking for a bird that is good with children and novice keepers.

Japanese Bantam

(Chabo)

Japanese Bantams are showy birds that are widely popular at poultry shows. Known in Japan as Chabo, they are thought to have existed in that country as long ago as the 1600s, based on depictions of them in paintings of the period. These are true bantams, with roosters reaching around 21 ounces (600 grams) and hens reaching 18 ounces (500 grams). They are characterized by their short stature and impressively tall tails, which are carried high over their heads. The unique genetics that give Japanese their special appearance also make them prone to poor hatch rates, and as a result they are notoriously difficult to breed. For keepers simply interested in raising them as pets or for show, however, they are ideal companions. They are friendly and easygoing, and thrive in smaller urban spaces. Japanese come in many varieties, some of which are extremely rare. The most popular colors are Black, Black-tailed White, Black-tailed Buff, and Birchen Gray.

FACTFILE
Purpose: Companionship, ornamental
Rarity: Uncommon, some varieties rare
Egg production: Good
Egg color: White
Origin: Japan
Size: Bantam
Temperament: Calm, affectionate

Above Japanese Bantams, or Chabos, are true bantams with tall tails and stunning feathers that come in a wide variety of colors. They are happy in small urban environments. **Opposite** In addition to the short stature and trademark tail, some Japanese Bantams can have the added variation of being a Frizzle. This is recognized as a trait in many parts of the world (and a breed in others), whereby the feathers grow outward and forward, creating a frilly appearance (see page 186).

Bantams & Ornamental

Serama

The diminutive Serama is the smallest breed of chicken currently available. Weighing in at less than 19 ounces (540 grams), and standing just 6 to 10 inches (15–25 centimeters) tall, it's the perfect pocket-size poultry for small backyards. The breed was developed from Japanese Bantams (see page 161) and small native bantams in Malaysia's Kelantan province in the 1970s, and it's relatively well known in that country, but it's so new in the United States and Europe that it's still considered rare. Besides its size, the Serama is known for its jaunty posture, with its chest thrown out, and head and tail held high. It comes in a wide variety of colors, most of which are not standardized yet, and which do not breed true from one generation to the next. Like its Japanese ancestor, the Serama is plagued with genetic problems that cause poor hatch rates in chicks, with the risk rising for smaller specimens. But Seramas are surprisingly good layers of tiny white eggs. They are also exceptionally friendly, especially toward humans, and are becoming popular as house pets. Seramas are ideal for small-scale chicken flocks, where they can be integrated with other small bantams. They don't need a lot of space, but they thrive on attention. They are beginner-friendly and good with children, although extra care should be taken due to their small size.

FACTFILE

Purpose: Companionship, ornamental
Rarity: Rare
Egg production: Good
Egg color: White
Origin: Malaysia
Size: Bantam
Temperament: Docile, affectionate

Above The colorful Serama has all the poise and power of a full-size chicken but in a tiny package. **Opposite** If you are considering a mixed flock, the Serama could be for you. Seramas are extremely friendly and love attention, plus they are happy to be part of a mixed flock. While this book can only encourage chickens as outdoor pets, many keepers find Seramas to be ideal house pets as well.

Hamburg

For backyard chicken keepers seeking to add a bit of flair to their flock, the Hamburg is an excellent choice. This is an ancient ornamental breed, with the penciled varieties originating in the Netherlands, and the spangled varieties being developed in England. The most commonly found Hamburg in the United States is the Silver Spangled—a flashy bird that sports a striking spangled pattern of black contrasting on white. Hamburgs have an erect stance, with tall tails, red rose combs, and large, dark eyes, and they also come in other varieties, including Golden Penciled and Silver Penciled, which are equally beautiful. While famous for its looks, the Hamburg is also a very good layer of medium-sized, white eggs. These are smaller birds, with roosters maturing to 5 pounds (2.3 kilograms), and females to 4 pounds (1.8 kilograms). Hamburgs are lively, active birds that free-range well, but they can be flighty, so are better suited to experienced chicken keepers.

FACTFILE
Purpose: Ornamental
Rarity: Uncommon
Egg production: Very good
Egg color: White
Origin: Netherlands and England
Size: Medium/light
Temperament: Active, alert

Above The Silver Spangled Hamburg (seen here) is both a good egg producer and a stunning bird. **Opposite** If you have a lot of space for free-range birds, the Hamburg could be for you, provided you have experience of chicken keeping and a set-up that can contend with active birds that are prone to taking flight.

Polish

One of the most unique-looking breeds of all, the Polish is about as flashy as a chicken can get. This ancient breed is characterized by the huge crest of feathers that covers the heads of both hens and roosters. Yet despite its fancy appearance, it was originally kept as a utilitarian, dual-purpose breed. The exact origins of the Polish are uncertain. It was standardized in the Netherlands in the sixteenth century, but descriptions of a chicken like the Polish occur in texts from Ancient Roman times, suggesting that its origins go much farther back still. Polish come in several recognized colors, including Golden, Silver, Blue, and Buff, depending on whether the variety is Bearded or Non-Bearded. This is a medium-sized fowl, with the roosters reaching up to 6 pounds (2.7 kilograms), and the hens up to 4.5 pounds (2 kilograms). Although primarily kept as an ornamental bird, the Polish retains most of its good egg-laying ability, and hens are dependable layers of large white eggs. Polish are prized as show birds and backyard pets. They are known to be very nervous and flighty, however, so may not be the best fit for keepers looking for more docile birds.

FACTFILE
Purpose: Companionship, ornamental
Rarity: Common
Egg production: Good
Egg color: White
Origin: Eastern Europe
Size: Medium
Temperament: Docile, affectionate

Above One look at this eccentric Muppet of the chicken world, and you would never guess this bird was originally kept as a dual-purpose utilitarian bird. This Tricolor Polish Bantam displays a beautiful range of colors, as well as the unique breed traits of small size and head crest.
Opposite This stunner struts across a free-range pasture, whether majestically or comically, it's in the eye of the beholder. This Golden Polish Bantam shares the same genetic trait of being a Frizzle, with feathers that grow in a different direction (see page 186). **Following spread** While Polish Bantams are known for their head crests, this mottled individual shows it off spectacularly.

Ayam Cemani

The Ayam Cemani is a stunning breed of chicken. Due to a genetic condition known as fibromelanosis (hyperpigmentation), every inch of it is pure black, including its skin, feathers, comb, and even internal organs. Its feathers are highly iridescent, with a turquoise sheen when exposed to sunlight. This results in an elegant, almost mystical-looking bird. Thanks to its unique appearance, the Ayam Cemani is one of modern poultry's most rare and sought-after breeds; well-bred Ayam Cemanis at one point carried a $2,000 price tag, and they continue to be significantly more expensive than most breeds today, particularly if they are of pure quality. Despite its very recent arrival on the Chicken Fancy scene, the Ayam Cemani is an ancient breed that has been steeped in mysticism in its native Indonesia since at least the twelfth century, historically used for cockfighting, exhibition, and occasionally religious ceremonies. Contrary to popular belief, the Ayam Cemani does not lay black eggs; they are actually a light cream. This is a medium-sized chicken, with roosters weighing up to 5.5 pounds (2.5 kilograms), and hens up to 4 pounds (1.8 kilograms). While highly prized for their looks, Ayam Cemanis tend to be flighty and independent, and often do poorly in very cold climates, so they're not considered the best breed for beginner keepers.

FACTFILE
Purpose: Ornamental
Rarity: Rare
Egg production: Fair
Egg color: Cream
Origin: Indonesia
Size: Medium
Temperament: Independent, flighty

Above This stunning Ayam Cemani rooster displays the green iridescence that can be seen on this breed's pure black feathers in sunlight. **Opposite** The Ayam Cemani is truly stunning, yet its all black body, feet, face, and comb cause it to look more like a work of art than a real, live bird. This breed is stunning and, while can be kept in warmer conditions, they are not a good first breed.

Egyptian Fayoumi

The Egyptian Fayoumi is another recent arrival to the United States and Europe with a very long history. This breed is believed to descend from the same landrace chickens kept by the ancient Egyptians along the Nile for their meat and eggs. This is a very beautiful breed with a uniquely exotic appearance. Roosters are mostly cream and silver, with a large black tail and black-and-silver barring on their rump, saddle, and legs. Hens have a silver head and neck, with a dark gray-and-silver, tightly barred body and wings. They have dark red single combs, and large dark eyes. Egyptian Fayoumis mature extremely quickly; hens may start laying as young as eighteen weeks, and roosters have been known to start crowing at five weeks or younger. They are lightweight birds, but not quite bantams. Roosters reach up to 5 pounds (2.3 kilograms), while hens reach up to 4 pounds (1.8 kilograms). Fayoumis are believed to have better disease resistance than Western breeds, and are currently under study for this ability. They are exceptionally heat-tolerant, and very capable when free-ranged on open pasture, but these tough birds are not very cold-hardy. The Egyptian Fayoumi is a great addition to ornamental flocks, but due to its flighty, wild nature, it may not be the best choice for beginners or keepers looking for a friendly breed.

FACTFILE
Purpose: Ornamental
Rarity: Rare outside Egypt
Egg production: Good
Egg color: Cream
Origin: Egypt
Size: Lightweight
Temperament: Independent, flighty

Above Fayoumis have a long history, but only a short time in the US and Europe. They are believed to be more disease resistant than their Western counterparts and can be very hardy (except in the cold).
Opposite This Fayoumi hen displays the tightly barred feathers of this breed and is seen here in a mixed flock. This breed can be independent and flighty, so it is best for environments designed to keep birds of this nature safe, and so is perhaps best for more experienced keepers.

Bantams & Ornamental

Yokohama

The beautiful Yokohama is the sort of breed that turns heads at poultry shows. Its ancestors were imported from Japan to Germany, where it was later developed as a distinct breed. While this breed comes in several color varieties, the Red Shouldered is the only color commonly found in the United States. Red Shouldered Yokohamas are very impressive birds with pure white feathers that contrast sharply with dark red wing patches on their shoulders. Their saddle and tail feathers are extremely long and drag along the ground behind them. Hens are nearly as stunning as the roosters, with a dappled red breast and wings, and a pure white neck and tail, which is carried high. Roosters reach around 5.5 pounds (2.5 kilograms), and hens reach about 4 pounds (1.8 kilograms). The Yokohama is a hardy bird with good heat tolerance, although its long tail feathers require extra care to keep clean for showing. Unlike many other ornamental breeds, the Yokohama is very docile and friendly, and is a great choice for beginners, backyard keepers, and families with children.

FACTFILE
Purpose: Ornamental
Rarity: Rare
Egg production: Fair
Egg color: Cream
Origin: Germany
Size: Medium
Temperament: Docile, affectionate

Above This highly ornamental breed is incredibly striking, yet its friendly and docile nature makes it suitable for beginner chicken keepers, even those with children. It is also tolerant to heat and surprisingly hardy. It does require extra care to keep its feathers clean, especially if the bird is also shown. **Opposite** This hen and cockerel pair beautifully display the dappled red body and long white feathers of the Yokohama breed.

Onagadori

The glamorous Onagadori is a rare ornamental breed from Japan. It carries a special gene that causes the roosters' tail feathers to grow perpetually, with no molting over the course of their lifetime. The result is a beautiful bird with ribbon-like tail feathers that can grow to a staggering 32 feet (9.8 meters) long! Despite recent awareness of the bird in the United States and Europe, the Ongadori is an ancient breed, believed to be more than four hundred years old. It has enjoyed special protection by the Japanese government for a long time, but still very few of these magnificent birds remain, and they are considered endangered. Despite some unscrupulous breeders' efforts to the contrary, true Ongadoris are not available outside the handful of breeders who maintain them. Those lucky enough to raise one will find this breed to have a lovely temperament and compact size, with roosters reaching just under 4 pounds (1.8 kilograms), and hens just under 3 pounds (1.4 kilograms). Ongadoris are poor egg-layers, and their long tail feathers require expert care to prevent damage.

FACTFILE
Purpose: Ornamental
Rarity: Very rare
Egg production: Poor
Egg color: Cream
Origin: Japan
Size: Small
Temperament: Friendly, curious

Above Another show-stunner, the Onagadori also originates from Japan. Unbelievably, the tail feathers never molt, making it the Rapunzel of chickens. **Opposite** This breed is now extremely rare and considered endangered. They are friendly birds that require a lot of care and attention to ensure the protection of their incredible tails. **Following spread** This long-tailed and glamorous Onagadori rooster in Thailand is a beautiful contrasted white and black.

Dong Tao

The enigmatic Dong Tao is a very rare breed that has only recently become known outside of its native Vietnam. Sometimes known as the Dragon Chicken, the Dong Tao is characterized by its massive feet, legs, and head, which give it a rather bizarrely swollen appearance; the feet are covered in red scales, which is what gives it a "dragon-like" look. The roosters are predominantly black with red hackle and saddle feathers, and a green-tinted black tail. The hens are significantly lighter, ranging from light cream to mottled brown. Both roosters and hens are very sparsely feathered. Once bred exclusively for royalty, the Dong Tao produces meat that is considered a delicacy. They are slow growers, but roosters at full size reach an imposing 13 pounds (5.9 kilograms), while hens tend to reach around 9.9 pounds (4.5 kilograms). The breed is notoriously difficult to reproduce, in part because of its enlarged legs. Dong Tao chickens are also rather delicate—their sparse feathering makes them very susceptible to changes in temperature, and their legs and feet require special care. Due to its rarity and slow growth rate, the Dong Tao is among the most expensive chicken breeds in the world, and if it becomes more widely available, it will be best suited to very experienced keepers who can dedicate the time and care it needs.

FACTFILE
Purpose: Ornamental, meat
Rarity: Very rare
Egg production: Poor
Egg color: White
Origin: Vietnam
Size: Medium
Temperament: Docile, affectionate

Above With the wonderful nickname of "Dragon Chicken," the Dong Tao has red scaled feet and legs, which are said to look like a dragon.
Opposite In spite of their sturdy legs and robust build, Dong Taos are delicate birds, due to their minimal plumage and unusual scaled legs. They require a lot of care.

Bantams & Ornamental

Sicilian Buttercup

The lovely Buttercup is a flashy ornamental breed with a unique comb. Instead of the single, rose, or walnut combs commonly found in chickens, Buttercups sport a cup-shaped comb that looks like two single combs growing together. This is a heat-tolerant breed that was first brought to North America by Sicilian immigrants in the nineteenth century, although it's believed to come from chickens that existed well before that, on the basis of chickens depicted in paintings from as far back as the sixteenth century. There were even depictions of chickens with buttercup-like combs found in ancient Egyptian tombs. Both males and females are a deep gold color with a lovely black tail. As they mature, hens develop small black spangles on their body, while roosters darken to a handsome copper-gold color, other than their tail, which is an iridescent black. Roosters reach just over 6.5 pounds (2.9 kilograms), and hens about 5 pounds (2.3 kilograms). These are alert and curious birds, able to free range well, and one of the most striking breeds available.

FACTFILE
Purpose: Ornamental
Rarity: Uncommon
Egg production: Good
Egg color: White
Origin: North Africa
Size: Medium
Temperament: Curious, friendly

Above With its highly unique cup-shaped comb and bright gold body, the Sicilian Buttercup is a stunning bird. As the female ages, she acquired black patches on her body, as seen here. **Opposite** With a temperament that is alert, curious, and friendly, this breed is a good choice for backyard keepers.

Campine

The Campine is a very old and very beautiful breed of chicken, originating in northern Belgium around the end of the nineteenth century, and developed almost concurrently in England. Campines come in two colors: Silver and Golden. Both varieties have narrow black barring everywhere on the body apart from the head and neck hackles, which contrasts well with their gold or white base color. They have large dark eyes, a red face, and blue earlobes. The result is a strikingly beautiful bird that is guaranteed to turn heads at local poultry shows. This is also one of only a few chicken breeds that features "hen-feathered" roosters, which lack the telltale saddle, tail, and hackle feathers that usually set them apart from the hens. The hens are decent layers of medium-sized white eggs, and rarely go broody. Campines are very active and intelligent, and do well when allowed to forage. They can be flighty, and therefore may not be the best choice for beginners or families with small children, but they're a great fit for keepers who wish to add a rare, fancy breed to their flock.

FACTFILE
Purpose: Ornamental
Rarity: Rare
Egg production: Good
Egg color: White
Origin: Belgium
Size: Medium
Temperament: Docile, affectionate

Above Campines come in two colorations: Gold (as soon above) and Silver (as seen opposite). Roosters appear more henlike than in other breeds, without the traditional saddle, tail, and hackle feathers.
Opposite If you have a large space for free-range roaming, as well as a set-up that allows for the safe-keeping of flighty birds, then the Campine could be for you. They are active and intelligent and can be docile and affectionate in the right environment.

Frizzle

The frizzle gene is a mutation that causes certain chickens' feathers to grow outward and forward, rather than flat and back, resulting in a frilly, windswept appearance. This mutation is recessive, and will not breed true, so these chickens remain uncommon. The Frizzle is classified as a breed in many countries, in addition to being a recognized trait, but in the United States it's only classified as a trait that occurs within other breeds—one that is highly sought-after by exhibition and companion chicken keepers. These chickens require some special care, as the frizzled feathers are not as insulating or waterproof as normal feathers. Breeds that are most likely to express the Frizzle gene are Japanese Bantams, Cochins, Polish, and Plymouth Rocks.

FACTFILE
Purpose: Ornamental
Rarity: Uncommon
Egg production: Varies
Egg color: Varies
Origin: Varies
Size: Varies
Temperament: Varies

Above Whether classified as a trait or a distinct breed, the Frizzle is the pug of the chicken world, so divided is the opinion.
Opposite Seen here looking like a pinecone, this Black Mottled Pekin Bantam is exhibiting the telltale feathers of a Frizzle.
Following spread Frizzles look perpetually windswept, and this feather type can require extra care.

CHAPTER TWO

CHICKEN KEEPING FOR BEGINNERS

THE WONDERFUL WORLD
OF CHICKEN KEEPING

If you are considering diving into the world of chicken keeping, congratulations! This is a truly rewarding hobby, and a rediscovered tradition among families around the world.

Chickens not only provide tasty, nutritious food in the form of freshly laid eggs, they are also delightfully entertaining creatures. In fact, many new keepers discover that watching their chickens amble around the yard is a relaxing, almost mesmerizing experience. These birds come in a dizzying array of patterns, sizes, and colors as well, ranging from standard backyard hens to rare, elegant birds. Some chickens may even become affectionate pets—their soft feathers and contended clucking can melt the hearts of children and adults alike.

Gathering together and feasting on backyard chicken eggs, sometimes laid just minutes beforehand, is a wonderful way to experience some sustainability and connection to your food. It also guarantees that your eggs come from a healthy, happy, and humanely raised flock. Children in particular delight in gathering their own basket of eggs, and they also benefit from learning about proper chicken care.

Thanks to their small size and relative ease of care, chickens are the perfect livestock to start out with. They thrive on both large farms and urban backyards, and are often the only livestock animal allowed in cities and suburban areas. A pleasant garden and small flock of chickens is all that it takes for an urban family to be able to experience the homesteading lifestyle.

Despite the many benefits of chicken keeping, however, this hobby is not for everyone. Chickens are messy and destructive, and they can live a long time. As outdoor animals, they are also prone to health and safety issues if not cared for properly. Before rushing out and buying a box of adorable fuzzy chicks, make sure you read this section of the book first, to familiarize yourself with everything that chicken keeping entails. But if you are fully prepared for your own flock, you'll find that becoming a chicken keeper—no matter the scale or purpose—is one of the most gratifying experiences you can have.

Opposite Chickens come in a stunning array of colors, patterns, and sizes—from bright white and deep black to all the colors of the rainbow at once, like this dazzling but sadly endangered breed from Austria, known as a Stoapiperl or Steinhendl.

Hens and roosters:
The birds and the bees

If you want to learn about the basics of animal reproduction, chickens make for an easy study. In fact, the incubation and hatching of chicken eggs has been used as an effective teaching experience in classrooms for generations. For those unfamiliar with how it works, from courtship to hatching, the process is outlined below.

Courtship

While not nearly as elaborate as the fancy dances performed by birds of paradise or the elaborate homebuilding demonstrated by bowerbirds, roosters—like the males of many other bird species—engage in a set of courtship displays to impress their flock of hens. They spend much of their time throughout the day crowing and proudly strutting around with puffed-out chests, showing off their impressive feathers and bright red combs. A rooster will also "tidbit"—offering tasty morsels of food to prove that they are a good provider (see page 203). All of these behaviors are meant to impress their hens, who are known to be choosy, preferring roosters who impress with their performance as much as their appearance. Unlike most bird species, roosters perform these courtship rituals throughout the year, though the behavior escalates in spring. There is one ritual in particular that they perform with their hens before mating—often referred to by chicken keepers as the "rooster dance." A rooster will begin by standing alongside his chosen hen, then he will drop his outer wing and circle her with a funny hopping motion, often while making a low clucking call. The dance ends when the hen squats down submissively and the rooster mounts her to mate.

Mating

Nearly always after a courtship display (and sometimes without any courtship), a rooster will select one of his hens and mount her as she squats submissively. Mating varies widely between different bird species, and for chickens the act is a bit tricky. Unlike ducks, for example, roosters do not have external sex organs (this is why sexing chicks is so difficult). Like their hens, roosters only have a single opening—the cloaca, which is used for both excrement and depositing sperm. Hens also lay eggs through the cloaca. Mating involves what is known as a "cloacal kiss"—a quick touching of the rooster's and hen's cloacae, during which the sperm is deposited. No penetration occurs. The act takes some balance work, and the rooster often has to hold onto his hen's neck feathers with his beak and shift his feet to get into position—an act known as "treading." The entire mating process only takes about five seconds. The rooster then dismounts, the hen fluffs her feathers, and both continue their day as though nothing has happened.

A hen can hold sperm inside her for up to two weeks, but a rooster will take no chances and will mate with his hens up to thirty times a day! Mating may be quick, but it can still be rough for the hen, especially if the rooster is much larger than her, or mates with her too frequently. If you keep roosters you must therefore keep a close eye out for any signs of injury—especially feather loss on a hen's back and wings. When this occurs, some keepers choose to separate them from the rooster; others will outfit their hens with a simple "hen saddle or apron" (see page 308) made from fabric to protect their back feathers from the rooster's treading.

Fertilization

Once the sperm enters her body, the hen actually has the option of ejecting it if she deems the rooster unworthy. Hens' criteria for gauging a rooster's suitability remain a mystery; the biggest rooster doesn't necessarily earn their favor. In general, though, hens seem to prefer roosters with large red combs that tidbit often. If the hen accepts the sperm, it may be stored in special pouches for up to two weeks, or it will move up the reproductive tract toward the infundibulum, a narrow section just outside the ovary where fertilization occurs. When a yolk is fertilized, a hen needs about 24 hours to grow and lay an egg. Then, once the egg is laid, her main job is to keep it warm.

Top left Amazingly, chicks are able to stand and walk within minutes of hatching. **Bottom left** The "rooster dance" is a courtship display involving the rooster encircling the hen with a dropped wing while hopping in a circle.

Brooding

When a hen is ready to sit on fertilized eggs, she exhibits a pattern of behavior called "brooding." When a hen goes broody, her appearance and behavior will radically change. The most obvious sign will be her refusal to leave the nesting box for many hours at a time. Broody hens are not to be messed with. They can become quite aggressive as they defend their clutch, and will growl and puff up angrily if approached. During this time a hen will stop laying eggs, instead focusing all of her energy on incubation, and she will rarely venture out to eat or drink. She will also lose all the feathers on her chest and underside, so her eggs get skin contact for better heat absorption.

Brooding is an instinct that can run strong or not at all, and it can vary greatly between individuals. Many breeds—commercial egg layers like Leghorns, for example—select against broodiness, so the hens of these breeds may never brood in their lifetime. Other breeds, such as Silkies, Cochins, and Orpingtons, are famous for going broody, and may exhibit this behavior frequently. The instinct can be so strong that a broody hen may steal the eggs of other hens, and if left to her own devices may sit on them indefinitely. Whether the eggs are fertilized or not makes no difference.

Brooding can be hard on a mother hen. She may become chronically dehydrated and undernourished, especially if the behavior goes on for longer than four weeks. Many chicken keepers are faced with having to "break" a broody hen so she can regain her health. While specific tactics vary, it boils down to two options: break the behavior by separating the hen from the nest, or give her some fertilized eggs or chicks to raise.

Interestingly, and very appealingly, a brooding hen will turn her eggs and cluck softly to them throughout the incubation period. As the chicks develop, they therefore learn the sound of their mother's voice, much like a human baby does. Chicks do not "imprint" on their mother like ducklings do, so they don't recognize her as the first thing they see. Instead, once hatched, they know her by her voice.

Opposite Contrary to popular belief, roosters will crow at any time of day. It is a territorial song they use to alert others that this patch is taken.

Eggs and chicks

The nutrients the embryo needs to grow are contained in a yellow yolk. Surrounding the yolk, a whitish section called the albumen contains both water and some additional nutrients. A thin white layer on the inside of the shell offers protection for the growing chick. A chick embryo develops quickly. By day 3, a tiny heart will be pumping inside the shell. Blood transfers nutrients from the yolk and albumen to the embryo. By day 5, the body is well on its way, with elbows, legs, and even eyes becoming discernable. By day 7, the beak and comb appear, and the embryo starts to move voluntarily. The beak will be present by day 10, and it will soon start to open and grow an "egg tooth." This is a temporary "tooth" on the top of the beak that the chick will later use to poke its way out of the shell. The first set of feathers (soft "down") will begin to emerge from bumps on the skin. By day 12, feathers cover most of the body, and the claws are fully formed. Between days 14 and 16, the chick will consume the albumen, gathering strength. Finally, on day 20, it will begin the process of hatching.

> *Shortly before they hatch, chicks start making little chirps or peeps to let their mother know they're on their way.*

There is an air pocket inside the shell, at the fatter end, and shortly before hatching the chick pokes into this pocket and takes a breath before starting to poke through the shell. Chicks don't then just magically pop out of their eggs. Using their egg tooth, they peck at the shell and make their way out gradually. They must stop and rest frequently, so the process can take between twelve and eighteen hours to complete—sometimes more.

Hatching starts with a phase known as "pipping." At this stage, the chick can be heard cheeping inside the egg. A tiny crack (or pip) will appear shortly after, which will grow slowly over many hours as the chick widens the hole with its egg tooth. Just before the final stage of hatching, a clear wide crack (or "zip") will appear in the egg. At this point the chick will take one final rest before pushing the two sides of the egg apart. It usually takes about twenty-four hours from first cracks to popping out. (And if you're wondering about the egg tooth, it falls off in the

Incubated chicks

Chicks that are raised to hatching in an incubator rather than under a broody hen spend the first four to six weeks in a brooder with a heat lamp to keep them warm (see page 243). Over this time, they'll grow feathers, just like the chicks who are keeping warm under their mothers.

Left and below Whether hatched and raised in an incubator or with their mother, chicks must be kept warm until their feathers are fully grown. **Opposite** Eggs come in a variety of colors, affecting only the outermost layer of the shell.

chick's early days.) It's tempting as a new keeper to help chicks as they hatch—the tiny baby birds are so clearly exhausted by the experience!—but the best thing is to let nature take its course. Chicks are surprisingly resilient, and a well-meaning keeper is more likely to injure them than help them.

Once hatched, the chick will consume any remaining yolk inside its egg. This can sustain it for up to 72 hours, if needed. Mother hens do not bring food back to the nest, as many other birds do. Instead, their chicks must be able to move around and forage for food with her. Incredibly, chicks are able to stand and walk about in just a few minutes, though those that experience any difficulty hatching can take up to a day. The chicks develop quickly, under the watchful eye of their mother. Her aggression will likely remain for several weeks; mother hens are extremely protective of their young. The chicks themselves dutifully follow her everywhere she goes, mimicking her movements and learning all the skills they need to survive. However, they need to keep warm until they grow their feathers, so they will nestle under her feathers throughout the day. By 8 to 12 weeks, they will look like miniature chickens— fully feathered and largely independent of their mother. At this point the males (called cockerels while young) are easily distinguished from the females (called pullets until they start laying). Maturity varies widely by breed, but most hens are capable of laying eggs at around 6 months old, though some breeds may start as early as 18 weeks.

Chicken eggs—myths and facts

When it comes to chicken eggs, myths and misinformation abound.

Myth 1: All chicken eggs can become chicks. False! Nearly all chicken eggs, including those bought at supermarkets, are infertile and unable to produce an embryo at all. This is because a rooster must be present to fertilize the eggs.

Myth 2. Hens need roosters to lay eggs. Also false! Roosters are necessary to produce fertilized eggs to hatch new chicks, but hens will lay eggs whether a rooster is present or not.

Myth 3. Brown eggs are more nutritious. False! Egg color only affects the outermost layer of the egg shell and has no impact on nutritional value.

Myth 4. All roosters are aggressive. Not always true. Many roosters can become very aggressive, but many others are exceptionally gentle and affectionate.

Previous Chickens can be cuddly, affectionate pets, especially if raised from chicks and breed chosen by temperament. **Above and below** Chickens display a wide range of behaviors, including foraging, grooming, and even tail-wagging!

Chicken behavior

Chickens are not known for their intelligence, but they exhibit surprisingly complex behaviors, all of which are necessary for these social, foraging birds to survive, just as they were for their wild ancestor, the junglefowl. Chickens can recognize up to one hundred unique faces across many species, from fellow flock members to their favorite humans, and even the family dog. They also make at least thirty distinct sounds; the noises they make, especially when chatting among themselves, are surprisingly nuanced. From contented clucking to angry growling and everything in between, every chicken sound has a meaning and every behavior has a purpose.

Social structure

Despite thousands of years of domestication, chickens' social structure remains very similar to that of the junglefowl. This social structure is steady and dependable, and is one reason why chickens were so readily adaptable to domestication in the first place. It allows them to find food safely, evade predators, and maintain a strict social order (what is colloquially known as the "pecking order").

Chickens live in flocks, much like many bird species around the world. These flocks, which naturally range in size from around five to fifteen birds outside of large commercial operations, ideally consist of one mature rooster per eight to twelve hens, along with a number of chicks and juveniles. Each flock holds a territory, whether it's the family backyard or a patch of jungle, and this territory is protected aggressively.

Roosters in particular are known to defend their home turf against any intruder, no matter the size or species, and their loud crowing is an announcement to all who live nearby that their patch is taken. This classic "cock-a-doodle-do" is one of the chicken's most unique attributes—no patch of countryside is quite complete without it. The behavior can start in male chicks as early as three weeks, but usually develops shortly before a cock hits maturity at about five months old. Contrary to popular belief, roosters will crow at any time of day, not just in the early morning. It is a territorial song that they use throughout daylight hours and occasionally even before the sun comes up. Roosters also tend to crow more often when excited, and especially if other roosters are nearby.

A rooster doesn't only announce and defend his flock's territory, he also enforces a strict social hierarchy within it. Each member has their place, with the most dominant birds getting first pick on food and resting areas. As long as the flock dynamic is healthy, this hierarchy is respected by every member of the flock, and fights rarely break out. On the rare occasion that squabbles do occur, it's the rooster who steps in and brings order. When a rooster is not present, a dominant hen may take up the leadership role to protect the flock's territory and establish the pecking order. She may even occasionally crow!

The ability of chickens—and especially hens—to maintain a healthy social dynamic without violence allows them to live in tightly packed groups, in very small spaces if needed, while remaining productive. Their incredible success as livestock is, in fact, driven by this sociability, along with their small size.

Daily routine

Chickens are very busy creatures. Throughout the day you'll notice them exploring, digging, bathing, preening, and clucking.

Foraging

A chicken may spend many hours a day foraging for food, even if feed is readily available. Because they are omnivores, their foraging behavior can range from grazing on grass, to digging, and even hunting. Chickens exhibit foraging behavior as early as a day old. They will peck the ground, scratching with their feet in search of buried morsels, and will readily investigate anything new in the vicinity in the hope that it might be edible. If an insect or small animal wanders too close, chickens enthusiastically give chase. They usually kill their prey quickly and will swallow it whole.

When foraging, chickens will occasionally share their food, in a fascinating behavior called "tidbitting." When the tidbitting chicken encounters a bit of food, they call out to beckon another to share it. The sound is a distinctly throaty cluck, which is uttered while picking up and tossing the tasty morsel to make it easier to see.

Mother hens tidbit to teach their chicks which food is best, while roosters tidbit to their hens as part of their ongoing courtship (see page 195).

Grooming

When they're not busy foraging, chickens spend ample time grooming themselves, including preening their feathers. Preening is a very important behavior for all birds, ensuring their feathers are in top condition, both to fly well and to keep themselves insulated against the elements. It involves using their beaks to clean and position their feathers, and spread a special "preen oil" over them—taken from a gland at the base of their tail. Chickens preen throughout the day.

Dust bathing is another very important grooming activity. Dust keeps chickens cool in summer, warm in winter, and clears their feathers of parasites and excess oils. To dust bathe, chickens will dig a depression in soil or dirt and lay down on their side. They will lie there, feathers puffed up, kicking dirt under their wings and close to the skin. It can look alarming at first, but this is very normal behavior that begins shortly after hatching. A healthy chicken will dust bathe at least once every day. Unlike other birds, such as backyard songbirds, chickens do not bathe in water.

Laying

Many people are surprised to learn that hens don't need a rooster to lay eggs. Egg laying is a natural part of hen biology, with or without a rooster. During her laying season, a hen will head off to a quiet, dark place where she feels secure to lay a single egg once every day or two. To the dismay of some backyard chicken keepers, a hen's chosen egg-laying nest may not be the nesting box provided, but hidden somewhere elsewhere in the backyard.

Laying hens can be very vocal, making that quintessential "bawk-bawk-ba-gawk" sound colloquially known as the "egg song." No one is entirely sure what it means. Hens seem to make the sound out of pride after laying an egg, and other hens often join in with egg songs of their own. Some chicken keepers speculate that the noise may be meant to draw predators away from the nest, or possibly to notify the rooster that they are done laying. No matter the reason, this is a very common behavior in most laying hens and can get quite loud if your birds are especially vocal. Like other chicken behaviors, the egg song varies by breed and individual.

Roosting

After a busy day of foraging, courting, grooming, laying, and making noise, chickens will dutifully line up to spend the night somewhere safe from predators in a behavior called "roosting." For backyard chickens, this means sleeping up on roosting bars inside a coop. For wild or feral chickens, it involves perching high in a tree, out of reach of most predators. Roosting is an instinct, so chickens generally don't need training to roost in their coop at night, although some birds may need a little extra help to learn where to go.

Other behaviors

Chickens also have many other minor behaviors that they may exhibit throughout the day, including:

Tail wagging

Chickens will occasionally "wag" their tails from side to side. They appear to do this when they feel especially happy or content. They may also wag their tails to adjust their feathers a bit. Hens often wag their tails after mating, after dust bathing, and after preening.

Crop adjusting

Before food heads into the stomach, it's held in a pouch called a "crop." Chickens occasionally adjust their crop—an action that looks very similar to yawning. They may do this several times in a row to move food down into their stomach or to get more comfortable. Excessive "yawning" over a longer period of time, however, can be the sign of a medical issue.

Shrieking

Chickens—and roosters especially—will make a high-pitched call or "shriek" when they spot a predator. They do this to warn the rest of the flock, letting them know that they need to take cover. The exact call chickens make can vary, depending on the type of predator.

Growling

Believe it or not, angry chickens do growl. The sound really is like a growl, but higher pitched, and a lot like how you'd imagine an angry dinosaur might sound. Hens tend to growl a lot when approached while laying, sitting on eggs (aka "brooding," see page 196), or while attending to chicks. Be careful if a hen growls at you—you will likely get an angry peck if you get too close.

Purring

Chickens can also make a series of low clucking sounds, similar to a purr. It's a noise they make when feeling especially content or affectionate. Very tame birds will purr when held by their favorite human.

Below Chickens can make at least thirty distinct sounds, including growling, purring, shrieking, and the quintessential "bawk-bawk-ba-gawk" egg song.

Problem behaviors

While the pecking order generally ensures harmony within a flock, occasionally things do go wrong, which is when chicken keepers need to step in to protect their birds.

Hen bullying

Bullying behavior in hens is a common issue with backyard flocks. When bullying occurs, one or more dominant birds relentlessly peck and attack more submissive birds, including other hens, newcomers, and young roosters. This goes far beyond establishing the pecking order, and can lead to injuries to a bird's face, head, and neck. In some cases, bullying can become so severe that the bird being picked on is denied food, water, and even access to the coop.

There are several tactics for nipping bullying in the bud. Arguably the most effective approach is to isolate the bully hen. This is as simple as removing the worst offender, and placing her in an isolated enclosure for a few days. The idea is that the bully will have to reestablish her place in the flock when

she returns, which can be enough to put an end to her bullying behavior. Another tactic that keepers try is the use of "peepers," which are small blinders worn on the chicken's beak that make it more difficult to peck other birds. They work in a similar way to the blinders worn by racehorses, blocking forward vision. This technique gets mixed results, but it can help in some circumstances.

As with most behavioral problems, prevention is key. Chickens that are kept busy, with lots of space to roam and plenty of opportunities to forage, are far less likely to develop bullying behavior.

Rooster aggression

Roosters have a very strong instinct to defend their territory. Unfortunately, this powerful instinct is often directed toward other chickens, humans (adults and children), and family pets. When aggression occurs, it usually involves chasing, pecking, posturing, and attacking with sharp spurs, which can cause substantial injuries.

Angry roosters can seem funny at first, but it's extremely important to address aggression as soon as it starts or it can escalate. There are many opinions on dealing with an aggressive rooster, and just as many strategies—from holding them down in a submissive position, to spraying them with a hose in extreme circumstances. Unfortunately, these tactics may only aggravate the issue. In some cases, rehoming or culling is the best option, especially if the rooster is particularly large, or when small children are present.

Contrary to popular belief, though, not all roosters are inherently aggressive. While there's no guarantee, you can take steps to avoid rooster aggression, starting with your breed selection. Several breeds are known for producing more docile roosters, including Brahmas, Orpingtons, and Cochins. Other breeds are known for developing more aggressive roosters, including production Rhode Island Reds and most sex links. In addition, roosters that are raised by hand and held frequently are far less likely to develop aggression toward humans later on. Selecting and raising your rooster from a day-old chick, including training them to come when called (see page 224) and picking them up daily, can make a huge difference.

Egg eating

It can come as a shock for new chicken keepers to find that their hens eat their own eggs soon after laying. Egg eating is relatively uncommon, but can be extremely frustrating to deal with when it occurs. This behavior sometimes crops up when birds are stressed, malnourished, or bored, but it can also be the result of a chicken's inherent opportunistic instincts. Egg eating may occur if eggs are frequently cracked or broken, too, which can happen if the nesting box is not adequately padded.

The best way to break an egg-eating habit is to stop it as soon as it starts—this is time-consuming but effective. You'll need to check on the nesting boxes regularly every hour or so, for a couple of weeks, removing the eggs as quickly as possible. You can also try the "dummy" approach, placing fake eggs in the nest. Both tactics will help the chickens lose interest in pecking and eating their eggs. If egg eating becomes too extensive to handle, rehoming or even culling the chicken may be necessary.

Behavioral problems can occur out of nowhere, but you can prevent most issues from the start by providing plenty of nutritious food, water, shelter, and a safe space for your birds to exercise and explore. A happy, busy chicken tends to be a well-behaved chicken.

Chickens that are hand-raised are more docile and happy to be around humans and family pets.

Health and support

Like any other pets or livestock, chickens can be plagued by a variety of injuries and diseases. In general, small backyard flocks experience few issues, thanks to their relative isolation, and most of these are also easily treatable at home.

Good chicken-keeping practices, including providing safe housing, fresh water, and balanced nutrition, can keep most ailments away; however, accidents and infections do occur. To avoid heartbreak, a backyard keeper must monitor their flock's health and behavior on a consistent basis. Being prey animals, chickens are very good at concealing illness or injury, so any sign of distress in a chicken therefore requires a rapid response. Preventing ailments also requires strict biosecurity measures to guard against fatal diseases.

Chickens recovering at home from injury or illness generally require similar treatment. Following the first-aid kit detailed on page 247, they should be isolated and kept safe and warm. In addition to treatment for their specific ailment, providing electrolytes and high-protein food can speed up recovery. Moistening their food is a great way to encourage a reluctant chicken to eat, and to keep them hydrated. Chickens are known to attack

members of their own flock if they appear ill, or are being reintroduced, so make sure to watch all interactions carefully as your chicken recovers.

While chickens generally don't receive routine veterinarian visits like other pets do, they sometimes require specialist care. Unfortunately, standard vets don't treat poultry or other birds, so it's a good idea to identify avian vets in your area well in advance—that way you'll know where to go in the case of severe illness or injury.

Chicken diseases

Chicken diseases can range from mild cold-like bugs to severe outbreaks of deadly disease. It's up to the backyard keeper to be acquainted with these diseases, and to be on the lookout for any symptoms in their flock. Most of these diseases are spread by wild birds, and especially other poultry, so to reduce exposure, backyard chickens should be kept away from wild birds as much as possible, and they should never come into direct contact with outside poultry until a 30-day quarantine is completed.

Keep in mind, too, that critter swaps, private sales, and poultry shows come with a higher risk of exposure to disease than registered poultry shows, which are strictly regulated. Never attend poultry events during a disease outbreak, and make sure to follow quarantine and other best practices if you or your birds are involved. Be extra wary if you choose to adopt or rescue birds from a commercial egg farm, where disease is often a major issue.

Biosecurity and quarantine protocol

A variety of biosecurity measures are key to maintaining a healthy flock, including always using a specific pair of shoes only for entering the chicken run, and disinfecting food and water containers regularly. Purchasing used coops and supplies can put your flock at risk, so avoid them if you can—or make sure you clean them thoroughly before use. If there's an outbreak in your area, you may need to keep your chickens in an enclosed, roofed run to prevent exposure to droppings or feathers from wild birds. And if you suspect an infectious disease is present in your flock, isolate any sick birds as soon as possible, and perform a deep clean of the coop, replacing all litter.

The premise of quarantine, when introducing any new chickens to your flock, is to act as if these new chickens are already infected with something, even if there are no symptoms present. The process entails a minimum of thirty days of zero cross-exposure between two or more groups of birds. This is a vital animal-husbandry practice, and disregarding it can have devastating consequences. When observing quarantine, newcomers must be kept at least 36 feet (12 meters) away from your existing birds. Food and water containers must never be swapped unless disinfected first. And the shoes and clothing worn by keepers should never be shared when caring for quarantined flocks.

Chicken > human diseases

When the general public thinks of chicken diseases, they are usually most concerned about zoonotic diseases: infectious diseases that can be passed from animals to humans. In fact, the US Centers for Disease Control and Prevention recently published a statement warning backyard chicken keepers against coming into contact with their birds, lest they get sick. Most members of the chicken-keeping community believe this concern to be grossly exaggerated, though; backyard flocks are significantly less likely to harbor disease than commercial flocks, and generations of keepers have managed to handle their birds frequently without getting sick at all. It's true that there is some risk of contracting a handful of diseases from chickens, but the vast majority can be avoided by observing basic hygiene.

Avoiding zoonotic diseases

To avoid the possibility of diseases being transmitted from your chickens to humans, remember to keep to the following guidelines.

- Always wash your hands well before and after handling your chickens. It's worth remembering that humans can pass diseases to their chickens, too.

- When holding your chickens, keep them away from your face—especially your eyes, mouth, and nose (which will also avoid any painful pecks near these sensitive areas).

- Wear a mask when cleaning the coop and run.

- Wash eggs before eating them.

- Instruct any children on how to handle birds safely.

- Restrict direct contact between chickens and infants, the elderly, and anyone who is immunocompromised.

Salmonella

Salmonella is a bacteria that infects the digestive system. It's commonly associated with contaminated food, and is often the reason behind mass food recalls. Salmonella is perhaps the most common cause of food poisoning in humans, resulting in symptoms including vomiting and diarrhea. It's rarely fatal, but it can cause severe illness. Salmonella usually clears up on its own in both chickens and humans, although it can require treatment with antibiotics.

Opposite and left The basics of good chicken keeping include balanced nutrition, fresh water, safe housing, and monitoring health.

Listeria

Similar to salmonella, listeria attacks the digestive system. Symptoms include vomiting, cramping, and diarrhea, along with muscle aches, confusion, and even convulsions in severe cases. It's a less common but potentially more dangerous disease than salmonella for both humans and chickens. It's treated with antibiotics, and in extreme cases it may require intensive care at a hospital.

Campylobacter jejuni

Bacterial infections caused by *Campylobacter jejuni* are relatively mild and asymptomatic in chickens, but they're a common problem with livestock and smaller captive birds. Campylobacteriosis is also a very common cause of food poisoning in humans, resulting in vomiting, diarrhea, and other forms of gastrointestinal distress. It usually clears up with a round of antibiotics, but it can cause severe illness, especially in immunocompromised individuals.

E. coli

Most animals, including humans and chickens, have a small population of *Escherichia coli* bacteria in their digestive system. When that bacterial population gets out of control, however, or a strain is introduced through infected food, illness can occur. Like salmonella and listeria, illness from *E. coli* is colloquially known as food poisoning. It's treated with antibiotics, although it can become life-threatening for the very young or elderly, in which case it requires hospitalization.

Histoplasmosis

This is a fungal disease that affects the respiratory system. It's contracted by inhaling the spores of a particular fungus—often via bird droppings. Interestingly, histoplasmosis does not affect chickens and other birds due to their high body temperature, but it can make humans ill, especially after repeated exposure. Symptoms are generally mild and flu-like, but infections can become severe in those with weakened immune systems. Histoplasmosis can be avoided by wearing a mask when cleaning out the chicken coop and run.

Avian influenza

The general public is now very familiar with avian influenza (or "bird flu"), due to a risk of some mutations presenting a danger to humans. Avian influenza can indeed infect humans, but it's primarily a bird-specific disease. As with any influenza, risk and severity of avian flu varies greatly by strain and year, but it's almost always extremely contagious, spreading quickly over large areas by migrating birds, or by unsuspecting humans carrying the virus from one area to another.

Due to its ease of spread, avian influenza can be a serious concern if severe strains appear. In early 2022, a particularly deadly strain was reported in North America. At the time of writing, it appears to present a relatively low danger to humans and other animals, but this strain has a reported 90 percent fatality rate in poultry, and any outbreak will require euthanasia for an entire flock. This strain is also spreading widely via migrating waterfowl, such as snow geese and ducks. Symptoms include dark purple discoloration in the legs and face, along with severe diarrhea and swelling around the eyes, but the most telling characteristic is sudden death, and it can spread rapidly through a flock. During outbreaks like this, the US Department of Agriculture and other government organizations release public warnings and ban poultry events. Zoos close their avian exhibits, birding events may be canceled, and commercial flocks that become infected must be culled onsite. Backyard chicken keepers don't necessarily need to panic during these outbreaks, but they should take extreme caution and monitor developments closely. If a severe outbreak is reported in your area, it's best to increase biosecurity measures—restricting exposed outdoor access if necessary—and report any suspected cases to the authorities.

Backyard chicken diseases
Marek's disease

This disease is caused by an avian-specific herpes virus. It causes tumors that, once they reach the nervous system, cause paralysis. Symptoms of Marek's disease include full or partial paralysis, with infected birds stumbling or dragging a limb when they lose the ability to move properly.

Marek's disease is highly contagious and nearly always fatal, spreading through the air and on the ground. Fortunately, it's 100 percent preventable with a vaccine, which is administered to day-old chicks at most commercial hatcheries.

Fowlpox

A viral disease that infects many bird species, fowlpox is transmitted through any open wound or via mosquitoes. It may present in one or both of two forms: cutaneous, which causes sores on the face; and diphtheritic, which causes sores in the mouth and upper respiratory tract. Otherwise healthy adult chickens usually overcome the disease, although the diphtheritic form is usually fatal, often due to starvation or asphyxiation from the sores. Fowlpox can be prevented with vaccines, but there's no treatment for it once it occurs. The best any keeper can do is keep their flock comfortable with plenty of fresh water, clean bedding, and high-protein feed.

Coccidiosis

Coccidiosis is a common disease experienced by backyard flocks, caused by a protozoa parasite. It attacks the intestine, causing bloody stools and anemia. This disease can be fatal in young animals, including chicks. Coccidiosis is most often treated preventatively in chickens through medicated chick feed. This doesn't completely prevent infection, but it will reduce severity and lead to immunity after exposure. Coccidiosis can also be treated using several over-the-counter drugs designed for use in poultry.

Respiratory infections

Respiratory infections can be the result of several specific diseases, including infectious coryza, Newcastle disease, and infectious laryngotracheitis, among others. Symptoms for these are very similar: nasal discharge, swollen eyes, and abnormal breathing. Diarrhea may also be present. Fatality and severity ranges widely, although most cases are relatively mild, and adult birds usually recover on their own within a couple of weeks. Antibiotics and fecal testing may be required, depending on severity. You can significantly reduce the chances of your birds contracting any respiratory illness by ensuring that the coop is well ventilated and the litter is kept clean.

External parasites

Chickens are prone to infestation by a number of external parasites, just as cats and dogs regularly suffer from fleas. The most common external parasites for chickens are feather mites, lice, and scaly leg mites. External parasites may appear at any time of year, in any climate, with some species being more common in different areas. Not only do these parasites cause discomfort, they can cause anemia, declining health, and even death if the infestation is severe enough.

Feather mites and lice

Symptoms of feather mites and lice include the presence of small mites around the vent area, as well as the feather shafts close to the skin, especially on the neck, legs, and head. Tiny mite droppings may also be seen, causing discoloration that is more obvious in birds with lighter-colored feathers. Lice especially can cause damage to the feathers, creating holes and giving them a thin, lacy appearance. Check your birds' feathers and vent for mites and lice frequently throughout the

year to catch any infestations early. If parasites are detected, treatment involves medicated poultry dust, which is available at most livestock stores. Be sure to follow the instructions precisely to clear up the infestation. Medicated shampoos and even injected medication are usually reserved for the most severe cases.

Prevention of mites and lice is not 100 percent guaranteed, but you can reduce the chances of an infestation by sealing the internal coop walls with paint, and keeping your chickens' litter clean. Avoid purchasing used coops and supplies, and always quarantine new birds to protect your flock from parasites. Chickens naturally rid themselves of mites and lice through dust bathing, so make sure your flock always has access to a clean and dry dust-bathing area.

Scaly leg mites

These mites cause the scales on chickens' legs to become thick and deformed, pointing outward instead of down, creating a rough appearance. These parasites are less likely to cause severe illness if caught early enough, but they are probably painful, and will lead to severe scarring. Treatment is relatively simple—apply petroleum jelly to your chickens' legs and toes daily for two to three weeks to smother and kill the mites. Spraying their legs first with an insecticide called permethrin can help speed up the process. Keep in mind that scaly mite damage takes a long time to heal.

Reproductive issues

Two of the most common afflictions for laying hens are egg binding and salpingitis.

Egg binding

Egg binding occurs when an egg takes longer than usual to pass through the oviduct and becomes stuck. When this happens, a hen will suddenly appear unwell and lethargic, sometimes seeming to strain or waddle. Usually, an egg-shaped lump can be felt around the vent area. Egg binding is extremely dangerous for a hen—if the egg is not removed it will kill her within 48 hours—so rapid treatment is key.

Effective treatment can usually be carried out at home. Bring the hen indoors, and place her in a tub of lukewarm water for a 10- to 20-minute soak. Epsom salt can be added to the water. Once soaked, rub olive oil around the vent area for lubrication and place her in a dark, warm place until she successfully lays the egg.

While any laying hen may experience egg binding at any time, some are more predisposed than others. Hens that start laying at a young age, or hens suffering from other ailments—including obesity, malnutrition, and stress—are more susceptible. Production-layer hybrids (see page 116) may also be more likely to suffer from egg binding. No matter what age or breed, make sure to keep treatment on-hand if you have laying hens, and inspect your flock regularly for signs of reproductive issues.

Salpingitis (lash egg)

Another reproductive issue is salpingitis, which is an infection of the oviduct. It's best known for resulting in a "lash egg." Lash eggs are not eggs at all, but infected masses of tissue that the hen's body produces and expels in an attempt to isolate and rid the body of infection. Salpingitis is not always fatal, but it is the number one cause of death in laying hens. If caught early enough, it can be treated with antibiotics, though the hen's laying ability will likely be compromised for the rest of her life.

Common injuries and ailments

In addition to the infections listed above, chickens can be prone to a range of other injuries and ailments, so it's good to be prepared for the most likely issues.

Shock

Chickens are extremely sensitive creatures, and they often go into a fragile state called shock in the aftermath of a traumatic experience, such as a predator attack. While in shock, a chicken may refuse to eat or drink. They will sit puffed, eyes half-closed, moving very little. Treatment for shock entails a 12-hour to several-day period in a dark, warm, quiet place with food and fresh water, where they can recover in peace. The chicken should be checked on periodically to monitor progress, but handling should be kept to a minimum if it appears to cause stress.

Bumblefoot

A common injury, especially in heavier breeds, bumblefoot is an isolated infection in the foot or toe padding, usually after an impact injury or small scratch. It appears as a lump or mass at the bottom of the foot, and causes discomfort and limping. To treat bumblefoot, clean the infected area well, and if you're confident enough to do so, carefully remove the abscess. Afterward, rinse the wound well with saline solution and pack it with an antiseptic cream, then wrap the foot to keep the wound clean while it heals. If you're feeling unsure, or if the bumblefoot keeps returning, an avian vet can provide treatment.

Frostbite

Chickens living in cold climates can be susceptible to frostbite, especially on their combs, wattles, and toes. Roosters and large-combed breeds are especially vulnerable. Frostbite appears as white or black patches, usually at the tips of the comb or wattles. As long as the case is mild, it's best to leave it alone—the dead skin will eventually fall off and be replaced with healthy new skin over time. In more severe cases, frostbite can cause permanent damage, and may require wound care to prevent infection. A properly winterized coop (see Heat and cold protection, page 226) can prevent or at least limit frostbite.

Broken beaks

Minor beak injuries are relatively common in backyard chickens. Cracks, chips, and breaks may occur due to fighting, collisions, or just bad luck while foraging. Fortunately, broken beaks will grow back, similar to nails. If you suspect one of your chickens has a broken beak, bring them inside for a close examination. If the injury is minor, no treatment is needed. More severe breaks, however, may be extremely painful and will interfere with a chicken's ability to eat or drink. It's possible to repair them with superglue and a soft-food diet, but these cases often require assistance from a vet.

Broken nails

Chickens spend much of their day scratching, so it's easy for a nail to get caught or broken. Unless the nail is actively bleeding, the injury is best left alone; the nail will grow back in time. If it's bleeding, or if the broken nail is still partially attached, it's best to bring the bird indoors to stop the bleeding and remove any broken bits with nail clippers. Make sure bleeding has stopped completely before reintroducing them to the flock.

Open wounds

Whether from predators, accidents, or fighting, chickens may find themselves with open wounds that require treatment. While chickens are very tough creatures, any open wound requires inspection and treatment. Cleaning even small wounds will prevent complications, such as infection or flystrike—a nasty condition caused by maggots burrowing into the skin. Wound treatment also prevents other chickens from making the wound worse by pecking.

As soon as you notice a wound, isolate the chicken and bring it inside for a quick assessment. If the wound is bleeding, help to stop it by applying cornstarch or styptic powder and a little pressure. Use saline solution to flush and clean the wound, making sure to remove any debris. Apply an antiseptic ointment and then wrap the area carefully with gauze and vet tape, if possible. Depending on the location and severity of the wound, the chicken may need to be isolated from the flock to prevent further injury.

Common chick ailments

Young chicks are susceptible to several ailments, most of which will be apparent shortly after hatching. Those listed below are the most common, and most are treatable, so long as the case is not too severe.

Cross beak

Also known as scissor beak, cross beak is a defect where the top beak does not line up properly with the bottom beak. It's most often genetic, but it can also result from vitamin deficiencies, difficulty in hatching, or improper incubation. Cases vary widely, from hardly noticeable to severe misalignment that prevents eating or drinking. Unfortunately, cross beak tends to get worse as a chick grows, and there's no treatment to reverse the condition. Some keepers opt to cull chicks with this condition to prevent suffering, but many other keepers are finding that with special care cross-beak chickens can live long and healthy lives. Depending on the nature and severity of the condition, you can provide extra care by implementing beak filing, increasing protein intake, or adjusting the position of the feeder and waterer.

Splay leg

Splay leg, or spraddle leg, is a common occurence in young chicks. Although it's sometimes caused by developmental issues while inside the egg, or occasionally from a vitamin deficiency, most cases of splay leg are the result of placing chicks on a slippery surface, especially during the first few days after hatching. Chicks with this condition will be unable to stand or move properly, due to one or both legs sliding out to the side. Fortunately, this condition is easy to treat if caught early. Any chick with splay leg should be separated from healthy chicks and outfitted with a brace. This can be made by threading an elastic band or hair tie through a piece of drinking straw, or made with vet wrap or even a Band-Aid or sticking plaster. The brace should fit comfortably and keep the chick's legs straight beneath them, about an inch apart. With the brace for treatment, a chick should fully recover from splay leg within one to seven days.

Wry neck

Sometimes called crook neck, or stargazing, wry neck is a condition that affects young chicks, though it can appear in adult birds as well. As the name suggests, the neck of a chick with this condition will have a twist in it, causing them to look perpetually upward or sideways and unable to hold their head upright. Wry neck may be caused by genetic issues or head trauma, but it's most often the result of a vitamin E deficiency. Chicks with wry neck must be separated from the flock for their own safety and treated with vitamin E and selenium supplements. They may also need gentle assistance with eating and drinking. With the supplements, a chick with wry neck should start to improve in as little as twenty-four hours, but it can take a month or more of treatment before they recover fully.

Pasty butt

This is a very common condition found in chicks, especially after they undergo any cold or stress, such as during shipping. This condition is exactly what it sounds like: the chick's vent area becomes pasted over with sticky feces. If left untreated, the excrement can completely block the vent, inhibiting the chick's ability to excrete properly. Pasty butt can kill delicate young chicks in a matter of days, so it's important to clean up every chick that experiences it. Make sure to use a warm, moist cloth and be careful to remove the droppings without injuring the chick's delicate skin. Pasty butt can be a symptom of a larger problem in the brooder, so prevent recurrences by ensuring the chicks are warm enough (use a thermometer to make sure) and that their water is always kept clean and at room temperature. Keeping the brooder in a quiet place away from animals and children can also prevent pasty butt.

Opposite Minor beak injuries are relatively common in backyard flocks, and broken beaks will grow back. In the case of a severe break, see an avian vet.

A safe space versus free-range

One of the first decisions that you'll have to make as a new backyard chicken keeper is whether to keep your chickens enclosed in a safe space during the day, or to let them free range or "wander." There are pros and cons to both approaches, but it primarily boils down to safety. Simply put, free ranging is riskier than predator-proof enclosures. For many chicken keepers, the trade-off of an active foraging flock is well worth the risk. For others, free ranging is too dangerous to consider. At the end of the day, keeping chickens in an enclosed space, letting them wander freely, or allowing a combination of the two is a deeply personal decision based on a range of factors. Fortunately, chickens can live long, happy lives in any arrangement, so long as the proper measures are put in place.

Living with predators

No living arrangement for a chicken flock should be made without careful consideration of local predators. Even in densely populated neighborhoods, chickens are at some risk of predation. They're especially vulnerable at night, when predators are most active, and when chickens' night blindness prevents them from protecting themselves. Animals that prey on chickens generally fall into three categories: aerial, ground-dwelling, and domestic.

Aerial predators

Aerial predators include hawks, eagles, and owls. These birds can strike without warning at any time of day or night. They are generally not strong animals, so netting or chicken wire tends to provide a sufficent barrier. Any open space, however, is at risk of an attack from the air.

Accipiters: The hawks in the genus *Accipiter* primarily prey on other birds. They include species such as the Cooper's hawk, sparrowhawk, and goshawk. Many of them thrive in urban environments and may be present at any time of year. Accipiters are fast, agile fliers. They can be bold, too, striking even when humans or dogs are nearby. The smaller species generally avoid attacking standard-sized hens, but they are a major threat to chicks, young pullets, and smaller bantam breeds.

Large hawks and eagles: Larger daytime birds of prey tend to be less common, but they are extremely effective predators. Including species like the golden eagle and red-tailed hawk, these powerful birds tend to frequent rural areas. Thanks to their size, they are more than capable of making off with a full-sized chicken. If one of these birds is present in your area, it's best to provide your flock with extra protection. Keep in mind, though, that it's illegal to harass or injure a bird of prey in numerous countries, including the United States and the UK.

Owls: Mid-to-large owl species like the great horned owl and barn owl are a moderate threat to backyard chickens. They prefer to strike at dawn and dusk, but attacks are known to occur during the day as well. Unlike accipiters, though, owls are more cautious when humans are present.

Ground-dwelling predators

Chickens are frequently the target of ground-dwelling predators. Strong, tall fences may be enough to keep some of these predators at bay, but most are capable of climbing and digging to get to their prey. Attacks from these animals tend to occur most often at dusk, night, and dawn, but there are often reports of bold attacks taking place in broad daylight.

Weasels: Weasels and other members of the genus *Mustela*, including minks and ferrets, are all enthusiastic chicken predators. Unlike their larger relatives, though, weasels are able to fit through very small gaps. They don't often consume chickens whole, but will kill every one that they can if given the opportunity. Weasels are also very fond of chicken eggs.

Rats and mice: These small rodents are not well known for attacking chickens (in fact, chickens often prey on mice) but eggs and young chicks are vulnerable. The presence of mice and rats may also

Careful consideration must be made of potential predators before starting a home flock. Predators can be aerial, ground-dwelling, or domestic. Suitable barriers, deterrents, and supervision are required to keep a home flock safe.

attract bigger predators, so it's best to remove them as soon as possible.

Foxes and coyotes: Foxes and coyotes are some of the most persistent chicken predators, with foxes being the top chicken predator in the UK. They are excellent diggers and extremely smart, capable of getting through most fencing. Many urban and suburban dwellers have foxes in their neighborhood, though they might not know it. When they attack, foxes and coyotes are more stealthy than dogs, often preferring to grab one or two chickens, then make off with them, so if a couple of chickens disappear without trace, it'll often be the work of a fox or coyote. Livestock guardian dogs are arguably the best deterrent against these predators.

Raccoons: These nighttime urban dwellers are frequent predators of backyard chickens in the United States. Like foxes, they're very intelligent, but having the advantage of hand-like paws, they're even more difficult to keep at bay. Raccoons are known for opening latches to coops and runs. If one is in the area, locks and carabiners may be necessary. Raccoons prefer to hunt at night and are generally shy, but they can be vicious if approached. Raccoons will kill chickens for fun, wiping out an entire flock in a night if the opportunity arises.

Badgers: Badgers are also known to prey on chickens. Like raccoons, they primarily hunt at night, but they're less likely to wipe out an entire flock in one go. Badgers are incredibly strong for their size, and can rip through weak chicken wire or plywood to reach their prey. These predators are expert diggers, too, so extra protection along the ground is needed if they're present in your area.

Wildcats and bears: Both wildcats and bears prey on chickens, although most urban chicken keepers need not worry about them. These predators may be less common, but because of their size and strength, they do require extra protective measures, such as electric wire and large livestock guardian dogs.

Domestic predators

Most backyard chicken keepers will have other pets at home, including dogs and cats. Unfortunately, these animals are predators and may present a threat to the flock. In some cases only a little training is required to avoid problems, but many pet owners will find that they have to take special precautions to keep their chickens safe.

Cats: Cats pose a very serious threat for chicks and smaller bantams. If you start with chicks at home, it's crucial to establish protection from cats, and to keep the birds in a separate, closed room to ensure they're not harassed. Once full grown, standard-sized chickens are usually too big and territorial to be threatened by a cat, but it's a good idea to watch their interactions closely, just to make sure.

Dogs: Feral and pet dogs are arguably the most common source of predation for backyard chickens. Like raccoons, dogs are known to kill for fun, and will decimate an entire flock in a matter of minutes if given the chance. If wandering dogs are present in your neighborhood, or if your own dog has a high prey-drive, it's recommended to use extra-strong fencing and other measures to protect your flock. Training pet dogs to accept the family chickens can be very effective, but supervision is still always recommended.

The advantages of enclosed spaces

Keeping chickens safe in an enclosed space is ideal for many chicken keepers and their flock. Peace of mind and neighborhood rules alone are reasons enough to keep chickens in a well-protected run, where they can still enjoy an active, enriched life.

Safety from predators

The primary reason for keeping chickens enclosed is to keep them safe from predators, but this only works if the chicken run is built to withstand attacks (see page 301). The run itself must be fully enclosed in wire—including the roof, to prevent aerial attacks by birds of prey. The construction material and wire must also be strong and secured well enough to withstand a charging dog or determined raccoon. Even the perimeter must be constructed so as to prevent predators from digging beneath the wire. It calls for a lot of effort and expense upfront, but predator-proof runs are the only way keepers can guarantee that their birds will not be brutally mauled or killed.

Flock control and health

If you're considering breeding or want to keep close tabs on your birds, then fully enclosed pens can be the way to go. In an enclosed space, you can maintain control over your flock's environment, from the food they eat to the breeds they mingle with. This level of control also prevents chickens from injury, poisoning, or exposure to disease.

Not all keepers choose to enclose their chickens in a run, however—usually for one or more of the reasons outlined in the next section.

The advantages of free ranging

It's true that keeping birds locked in a run results in a lack of freedom and enrichment, and this is a compromise that many keepers struggle with. Chickens that are kept in runs throughout their lives—particularly if those runs are too small or crowded—are more likely to exhibit behavioral issues. As a result, free ranging is still a very popular method of keeping chickens, where they are allowed to wander the homestead, farm, or village, foraging for food and enjoying an active life.

Health and happiness

With plenty of space to explore, a free-ranging chicken is never bored. They can spend most of the day foraging for insects, worms, and grasses, or dust bathing in a bare patch of dirt. When free ranging is allowed, chickens are less susceptible to common behavioral problems like bullying, overbreeding, feather picking, and egg eating. Even when bullying does occur, a hen can easily escape to avoid conflict.

It's hard to argue that chickens don't display happiness when let out to free range. They will tend to dash out all at once, flapping and running about the yard excitedly. For many backyard keepers, offering this kind of freedom is a high priority.

Better eggs for less feed

Thanks to their healthy, varied outside diet, free-range hens require less commercial feed and produce more nutritious eggs. On average, the egg of a hen with access to pasture contains twice the amount of vitamin E, 38 percent more vitamin A, and more than twice the amount of Omega-3 fatty acids.

The disadvantages of free ranging

Of course there are also numerous downsides to free ranging. The most obvious drawback is predation, but other factors also need to be taken into account.

Predators

Predators are always a fact of life when raising chickens, but much more so if you allow them to free range. So many other creatures delight in hunting and killing chickens, and their attacks can be brutal, occuring at any time of day without warning. Sadly, the results are often fatal. For keepers raising chickens for profit or food, any predator attack is a loss. For families who raise chickens as pets, losing birds to predators can be outright traumatic. Even chickens that survive will suffer from the stress of an attack; they might experience shock, refuse to eat, or stop laying altogether.

Escapees

Free ranging safely is just as much about keeping chickens in as it is about keeping predators out. Because they can be so destructive and messy, chickens are often unwanted guests on neighbors' properties. And a hop over a neighbor's fence can result in tragedy if those neighbors keep dogs.

Food thieves

Once songbirds and rodents get wind that a free buffet of chicken feed is available, they will gladly take advantage. While this doesn't necessarily pose a danger to chickens, it does lead to pest issues and a big spike in the feed bill.

Destruction

Chickens are surprisingly destructive creatures, and they show very little regard for a homeowner's possessions or space. Given a few hours to freely

simply too high, even in the most urban areas, and chickens are defenseless in the dark, unable to see well enough to escape.

Enrichment tips for chicken runs

If your backyard is simply too dangerous or difficult to allow chickens to free range, there are several activities that will help provide the benefits and freedom of free ranging safely.

Build chicken tunnels

Getting creative with different enclosures will allow your chickens to enjoy a bit of pasture throughout your property while still keeping them enclosed during the day. Chicken tunnels are one option. These consist of a simple wired passageway that your flock can use to get from one enclosed space to another, usually along a fence line. This allows them to get plenty of exercise while enjoying safety from hawks and owls.

Use portable tractors

Chicken tractors are a tried-and-true method of keeping chickens safe while out on pasture. The tractor is usually a sturdy A-frame structure that moves on wheels. It can be moved around every day or so, allowing the chickens inside it to enjoy fresh grass and bugs just as they would if they were free ranging. For smaller backyards, a lighter-weight chicken tractor can be used to give a flock access to different areas in the yard during the day.

Provide enrichment

Chickens are busy creatures that get bored easily. Fend off boredom and bad behaviors by providing a few free-range activities inside the run. Bring in tree branches from outside and hang them up for new perching opportunities, or rake in a fresh pile of leaves or grass trimmings for the birds to peck and scratch through. You can also hang fresh herbs, veggies, and fruit throughout the run, challenging the birds to jump and peck for a tasty treat. Done consistently, these little activities and treats can help boost your flock's nutrition and egg quality just as well as if they free ranged on pasture.

forage on a property, chickens will happily uproot flowers, dig up a lawn, and leave piles of droppings on back porches. If you prefer a tidy, well-kept backyard, free-ranging chickens will cause you a lot of frustration.

A happy medium?

Free ranging versus safe space doesn't have to be an all-or-nothing decision. Many chicken keepers opt for a compromise, with a primary safe run alongside some limited free-ranging time for their flock under close supervision. Still others find compromise by bringing some free-range activities and benefits to their run-kept chickens, so the flock can enjoy active foraging without the risk.

Keep in mind, though, that no matter which living arrangement you choose for your birds during the day, a securely closed coop at night is a necessity. The risk of predator attacks overnight is

Above For many keepers, a mix of an enclosed run alongside limited free-ranging time to be the best solution for keeping chickens safe and happy.

Free ranging safety tips

Free ranging does not necessarily mean leaving your birds unprotected. There are several tried-and-true strategies to keep free-ranging chickens as safe as possible.

Limit contact with wild birds

If you choose to free range, consider removing bird feeders from the area. This prevents the spread of disease, like fowlpox and bird flu, which can easily transfer between wild birds and chickens (see Health and Support, pages 208–15).

Provide hiding places

Hawks and owls are a frequent issue with free ranging, including in urban areas. You can help protect your flock by offering cover, like bird netting or bushes to hide in. This doesn't guarantee safety, but it can buy your birds a little extra time to get away if they come under attack.

Provide supervision

Leveraging dogs to provide "guard duty" while your chickens are free ranging is a great way to deter most predators, and has been used for centuries on farms all over the world. Not all dogs make good livestock guardians. If you have a dog already, consider hiring a trainer to see if they're up to the task; if you're looking to purchase a dog, consider one that has been bred for the purpose of guarding livestock. If dogs are not an option, humans are even better! Many backyard chicken keepers opt to free range their flock during restricted hours, while they or their children are outside with them. Even bold predators will usually stay a safe distance away if humans are present.

Enclose a larger space

Backyard chickens are usually in a fenced-in, open backyard to wander. This protects them from most ground predators and keeps them within sight. While not entirely risk-free, enclosing a larger space works well for many, and is much more affordable than building a large enclosed run.

Use deterrents

You can make predators a lot more uncomfortable around your property by employing a number of deterrents. A big fake owl that's moved around frequently can help keep birds like hawks and owls away. And mountain-lion urine can make animals like coyotes too nervous to linger.

Train to come when called

Chickens will do just about anything for their favorite treat, and this makes them easy to train. A chicken that will come when called (see box below) is not only adorable—it also makes life much more convenient, and helps to keep them out of trouble (such as when they fly on the roof or escape the yard).

Use the 2-barrier rule

If you have a dog that poses a danger to your chickens, always use the "2-barrier rule" to prevent accidents. The idea is to maintain two barriers (gate, door, fence, etc.) between the chickens and the dog. That way, if the dog breaks through one barrier, or is let out by mistake, there is still another layer of protection for the birds.

3 steps to train your chickens

Training chickens to come when called is remarkably simple, and best done when your birds are as young as possible.

1. Hold out your chickens' favorite treat (dried mealworms and other larvae are a popular choice). Get their attention by tossing some on the ground.
2. Make a calling sound, like a whistle or a repetitive "chick-chick-chick-chick!"
3. Give plenty of treats as your birds come rushing over.

Now teach your chickens to follow you by continuing Steps 1 to 3 as you slowly walk around.

Above Chickens can be surprisingly destructive if allowed to free range anywhere, so plan your outdoor space accordingly. **Below** Dogs and chickens can be trained to be excellent companions. However, it is advisable to always supervise interactions.

Heat and cold protection

Chickens are remarkably hardy creatures, able to withstand both extreme heat and cold, as long as they have adequate shelter and care. The temperatures that a chicken can handle will generally depend on their breed. All chickens are believed to descend from the junglefowl, a bird native to Southeast Asia and India. This bird takes heat and humidity in its stride, and its descendants that were bred for warmer climates share its impressive heat tolerance. Other chickens, however, have been bred for thousands of years to withstand much colder temperatures. They lack the junglefowl's capacity for heat, but with adequate protection, these cold-hardy breeds can easily survive harsh, snowy winters.

For the easiest and best results, the chicken breeds you select should match your climate. Cold-tolerant breeds, like Cochins, Brahmas, and Orpingtons, for example, will thrive throughout cold winters without the need for any special care. Heat-tolerant breeds, like Sumatras, Sebrights, and Leghorns, will perform well through the heat of summer, with a much lower risk of heatstroke.

Having identified the most suitable breeds for your home flock, adequate heat and cold protection then comes down to proper housing.

Heat protection
In order to keep a flock healthy and comfortable on hot summer days, a chicken coop and run should include plenty of shade and fresh water at all times. Chickens depend on ground cooled by shade to survive the midday heat, so make sure they always have an area protected from the sun, especially during the hottest part of the day—the larger the better. If your chicken enclosure doesn't get much shade in summer, consider planting dense trees or shrubs nearby, or simply hang up shade cloth positioned on the south and west sides of the run (if you're based in the northern hemisphere) to keep the direct midday sun at bay. It's also advisable to construct a sheltered dust-bathing area, protected from the sun and rain. Keep in mind that a single chicken can drink up to a pint (almost half a liter) of water a day in summer. They need to stay hydrated to survive hot temperatures, so always keep waterers filled up and out of direct sunlight.

Other tactics that can help chickens stay comfortable in the heat include providing tasty frozen fruits and vegetables. Frozen peas, beans, and broccoli are excellent choices, as are frozen berries and even chilled Greek yogurt. On really hot days, you can also spray down a shaded patch of dirt or sand so that the chickens can use it to dust bathe and cool off. Some keepers even set up a mister system in the run to help their birds stay cool. All of these techniques will not only ward off heatstroke, they may also result in better egg laying by keeping the birds less stressed.

Cold protection
In very cold climates, a flock needs protection from the elements to stay healthy and comfortable, and the primary way to do this is by providing adequate shelter.

A proper winterized chicken coop should first ensure draft protection. Coops that allow drafts in will chill the air and prevent the chickens from staying warm. This is especially important on cold and windy nights. One easy way to identify potential drafts is to

Above While this free range blue Cochin hen seems unperturbed by the deep snow, for many chickens, it is advisable to dig out paths for them to walk through and essential for all to have a weatherproof place to roost.
Opposite This stylish chicken is kept cozy in the arms of the book's author, while bedecked in a handmade hat and scarf (purely for decorative purposes).

go inside the coop during the day and shut the door. Anywhere that lets light through will require extra protection. To reduce drafts, you can use weather stripping (just make sure your chickens can't peck or eat holes in it), or cover the sides of the coop with a tarp or heavy blanket (as long as the ventilation is not covered; see below). Covering open windows with clear plastic or bubble wrap will also prevent drafts, in addition to providing a bit of insulation from the cold.

Another very important, yet frequently overlooked, aspect of a winterized coop is adequate ventilation. If the coop is not well ventilated, moisture and ammonia can quickly build up. Too much of this excess humidity can lead to respiratory illnesses, and cause moisture to condense on the birds' combs. Once their combs are exposed to the outside air, they're then prone to frostbite (see page 214). The purpose of ventilation is to keep the coop as dry as possible and allow for airflow high up above the roosting chickens. The best place to install ventilation is at least 24 inches (around 60 centimeters) above where they roost. The more crowded a chicken coop is, the larger the ventilation will need to be.

It's worth noting that the ambient humidity outside does not result in a greater likelihood of frostbite inside the coop. The key is to keep the coop's humidity as close to the humidity outside as possible. In addition to proper ventilation, you should keep water outside the coop at all times.

Additional cold-survival tips

As well as a winterized coop, there are many tried-and-true tricks to help chickens thrive through the winter. From fall onward, offer them extra protein-rich foods to encourage the growth of healthy, thick plumage. On very cold days, give them some extra scratch grains or corn right before bedtime; the extra calories will help keep them warm through the night.

To make the coop itself more comfortable, many experts recommend filling the floor with an extra-thick layer of bedding, and installing extra-wide perches so the birds can keep their feet warm while roosting. You can also install insulation inside the coop, but make sure the chickens can't access it or they'll likely to peck at it and eat it. To encourage them to explore outside the coop on snowy days, some extra protection around the outdoor enclosure can also

be very helpful. Try covering the run walls with clear plastic sheets, tarps, or even straw bales to keep out the snow and wind, and provide them with piles of hay and straw to encourage foraging. This promotes an active lifestyle, prevents boredom, and may even reduce bullying. Chickens generally don't like walking through snow, so if your run doesn't have a roof to keep snow out, make sure to shovel walkways so that they can get outside.

To heat or not to heat?

Heating a coop during winter is a hotly debated topic in the chicken-keeping world. Heaters can keep birds more comfortable, especially in extreme weather, but they come with some risk, too. Fire is one such risk. Every year, chicken coops and barns tragically catch fire from heat lamps. The risk can be mitigated, but it's not 100 percent avoidable. Power outages are another concern. Chickens that come to depend on heat at night may not develop the warm, thick feathers they need to withstand cold temperatures, and a sudden power outage during a blizzard could shock and kill them. Too much heat inside a coop can also cause warm moisture to condense. And when the birds are exposed to the much colder air outside, the sharp drop in temperature can shock their system, cause frostbite, and even lead to pneumonia.

Supplementary heat is sometimes necessary for the well-being of chickens, though, despite the risks, and chicken keepers will rightfully want to make sure their birds are comfortable. Reasons to consider using heat in the coop include during heavy molts in spring and fall, when chickens lose most of their feathers and are most vulnerable to extreme cold, or when a chicken is recovering from an illness or injury. Other times when external heat may be required include extreme, sudden temperature drops, especially early or late in the season, and when raising cold-susceptible breeds or young birds that aren't yet acclimated to the cold.

Remember, when it comes to electricity and heat in a chicken coop, safety is paramount. And no amount of supplemental heat can replace a well-constructed and draft-free coop.

Nutrition

New chicken keepers will quickly realize how many options there are for feeding chickens. Many keepers joke that these birds are essentially pigs with feathers, as they will eat just about anything. Whether you are establishing a homestead or simply keeping chickens for fun, offering them a varied, nutritious diet is an excellent way to ensure they'll live long and healthy lives.

How chickens eat and drink

Like all birds, chickens have a digestive system that is very different to that found in mammals like cats and dogs, so for proper care, it's important to understand how this works.

The differences are apparent right away, starting with the beak and teeth. Chickens use their lightweight beaks to pierce and bite food. They don't have a developed tongue or throat like mammals do, so when drinking they must scoop up water with their beaks, then tilt their heads back to let gravity draw the water down. Because they lack teeth and cannot chew, they must consume bite-sized chunks of food; when able, they'll simply peck off smaller pieces, otherwise they'll grasp food with their beaks and shake their heads to break off smaller pieces. Once the morsels are small enough, chickens swallow them whole.

Once swallowed, the food moves into the *crop*, which acts as a sort of holding pouch. The crop is located between the chest and throat. In the morning, before breakfast, it's hard to see, but it fills up throughout the day, especially after a large meal, appearing as a soft lump or bulge. After being held in the crop for up to twelve hours, food is then moved toward the *gizzard* in a slow process that is driven by tiny muscles in the crop. The gizzard is not quite a stomach, but it serves a similar purpose. Here, the food is further broken down with digestive enzymes and small rocks known as *grit*. Starting from as young as a few days old, chickens (and all birds) consume grit throughout their lives. This is nothing to be concerned about—in fact, it's a healthy part of their diet and should always be accessible to them. Due to a bird's lack of teeth, the grit is extremely important for breaking down food. For chickens kept without access to rocky soil, a commercial grit supplement may be provided instead.

After the gizzard, food is moved and absorbed by the small intestine, ceca, and large intestine. The final stage before excrement is the cloaca. This is where the urinary, digestive, and reproductive systems intersect. Chickens and other birds do not have bladders and do not urinate. Instead all solid and liquid waste is excreted together out of the cloaca and through the vent.

Water

Like all animals, chickens always need access to fresh, clean water. A single hen consumes an average of 1 pint (just over half a liter) of water a day in hot weather, but they need plenty of water in winter, too, so it's important that they can access water at all times, no matter the weather.

The way water is provided depends entirely on the keeper's preference. During summer and in milder climates, water bottles or buckets with nipples that the chickens peck provide a handy and clean way to offer fresh water. Other keepers may prefer a simple shallow dish or vertical waterer. During winter, most keepers opt for a heated dog bowl or a special heating plate to keep water from freezing. Drowning is a real concern for all chickens, and especially chicks, so they should never have access to deep, open water, such as a pond or an open bucket.

Commercial feeds

No matter the purpose of keeping chickens, they should always be provided with standard commercial feed throughout the day. All store-bought feed should come with a nutrition panel, and be formulated to ensure every member of the flock receives adequate nutrition for every stage in their lives. There are many options available for commercial feed, ranging from big-name brands to locally milled and specialty brands. Commercial or store-bought feed for any chicken generally contains grains such as corn, barley, and wheat, along with soy or cricket meal for protein, in addition to vitamins and minerals. As your flock grows, you'll need to ensure that you select the right commercial feed for its specific needs.

Chick starter/grower

Chick starter and grower feed should be the default for all young chicken flocks up until they're ready to lay. These blends are higher in protein to promote healthy growth and feather development. Chick starter is available in two forms: medicated and non-medicated. The medicated feed contains an active ingredient used to treat and prevent coccidiosis (see page 211), a disease that's very common in unvaccinated chicks. Choosing whether or not to use medicated feed is a personal decision and a hotly debated topic in the chicken-keeping world. Many keepers prefer to use it to prevent illness while the chicks are young and vulnerable; others opt to treat only when necessary.

Flock raiser

For meat broilers and chickens in mixed flocks with ducks or other fowl, a flock-raiser feed is a better choice. It contains extra niacin, which is needed for fowl like ducks. It also contains even more calories and protein than chick starter, and is perfectly healthy for chicks and chickens of all ages. Chick starter or flock raiser should be the default feed for all non-laying chickens.

Layer feed

Once she begins laying, a hen's nutrient requirements begin to change, and therefore so should her feed. At this stage, she needs less protein and more calcium to support healthy egg and reproductive development. It's important to note that only laying hens should have access to layer feed—the high calcium content, which usually comes from added calcium carbonate, can cause liver damage and other issues for chicks and roosters. If your flock includes a mix of ages and sexes, it's best to offer only flock raiser or chick starter, then provide extra calcium supplements in a separate container. The laying hens will naturally be drawn to the supplements, while the chicks and roosters will largely ignore it.

Scratch grains

Scratch grains are a common treat for flocks of chickens, usually containing cracked corn, wheat,

barley, and oats. They are also very affordable and easy to buy in bulk. Keep in mind, however, that scratch grains are a calorie-rich supplement and not a substitute for a bag of nutrient-balanced feed. Feed a handful to your chickens every night before bedtime, especially in winter. The extra calories will help keep them warm and full throughout the night.

Supplements

There are many nutritional supplements available for chickens, some of which are crucial for a healthy flock.

Grit

As mentioned on page 232, all birds require grit, which is simply small pebbles or rocks that they consume to help their gizzards digest food properly. Chickens that are fed a solely commercial diet may not need grit, but any chickens with access to bugs, leaves, and other whole foods must have access to grit on a regular basis. Free-range chickens generally find grit on their own, but birds kept in an enclosed run should be kept provided with it throughout the year. Grit is generally easy to find in feed stores. Most keepers prefer to offer it freely in a small container, so chickens can help themselves if they need it.

Calcium

Calcium supplements should be offered to laying hens to ensure strong eggshells. This is especially important if the hens are not fed commercial layer feed. The most common calcium supplement comes in the form of crushed oyster shells, which is easily found at most feed stores. Many chicken keepers, however, prefer to feed hens with their own eggshells. This replenishes their system, is sustainable, and hens often seem to prefer it over oyster shells. To do this, simply rinse the shells and lay them out to dry; you can microwave them or heat them in the oven to speed up the process. Once dry, simply crush the shells into small pieces, then store them in a dry container. Offer them to your chickens as you would oyster shells. The hens will consume them as needed to stay healthy.

Herbs

Herbs are a tried-and-true supplement that may be offered to chickens. They can be dried and sprinkled onto their food, or hung up fresh for the birds to peck at throughout the day. Besides their potential health benefits, herbs provide variety to a chicken's diet, and the birds seem to genuinely enjoy grazing on them.

Marigold

The dried petals of marigolds (a group of flowers belonging to the genus *Calendula*) are sometimes fed as a supplement to laying hens to improve the vibrancy of their yolks. Marigold is also thought to offer respiratory and anti-inflammatory support.

Oregano

Oregano is a popular herbal supplement for backyard chickens and has proven effective for reducing common poultry illnesses like coccidiosis and bronchitis. Oregano helps support the immune system, and may have additional anti-parasitic and anti-inflammatory properties.

Garlic

Garlic is also commonly used as a supplement for backyard chickens due to its reported immune-boosting properties. Many keepers also swear by its ability to help induce egg laying. It can be given as a powder or freshly chopped.

Apple cider vinegar

Apple cider vinegar (ACV) is a popular, if occasionally controversial, supplement that can be added to chickens' water. Many experienced keepers swear by the health benefits of ACV, while others insist it has no effect at all. ACV is certainly helpful for keeping water cleaner, and it can help reduce slime and other contaminants that may cause disease. Recent studies also show that it may support healthy digestion and improve calcium absorption, as long as it's given in moderation. To provide ACV as a supplement, add 1 tablespoon of vinegar to 1 gallon (4.5 liters) of water.

Kitchen scraps

It's no secret that chickens will eat almost anything. Besides grain and grass, they are known to eat mice, snakes, birds, and even carrion on occasion. They also have a fondness for human food, making them excellent for reducing waste. As long as the food offered is healthy, a backyard flock will thrive with regular treats of kitchen scraps.

Kitchen leftovers make fun treats, and keepers quickly learn which are their flock's favorites. If you want to reduce food waste and supplement your chickens' diet with leftovers, consider keeping a washable "chicken scrap" bucket in your kitchen and fill it up throughout the day with things like bread crusts, carrot tops, and leftover salad. The chickens will enjoy the variety, and their eggs will be even more nutritious!

Please note that in some countries, such as the UK and countries within the European Union, feeding kitchen scraps to farmed animals—including pet chickens—is illegal. This is to avoid the outbreak of diseases like swine fever and hand-foot-and-mouth. The ban has been in place since 2001 in the UK, after a terrible outbreak that was connected to leftover food being fed to livestock. At the time of writing, the US has no such ban.

Dos and don'ts

Chickens will happily devour just about anything you offer them, so it's important that they're only fed healthy food, and that kitchen scraps make up only about 10 percent of their diet for balanced nutrition.

Common household foods that can be enjoyed by your flock include:

- Apples
- Stone fruits, like cherries and apricots
- Raspberries, strawberries, and blueberries
- Bananas
- Melons
- Unsweetened cereal
- Cooked oatmeal (porridge)
- Cooked root vegetables
- Raw salad greens, like lettuce and spinach
- Cooked fish, steak, and other meat (give the chickens the bones and they will pick them clean)
- Greek yogurt in limited amounts (chickens are lactose-intolerant)
- Summer and winter squash
- Popcorn
- Cooked pasta
- Raisins and grapes

The following foods are considered toxic for chickens, or at least extremely unhealthy, and should never be provided:

- Potato skins and sprouts
- Avocado skins and seeds
- Chocolate
- Alcohol
- Coffee beans and grounds
- Onions
- Rhubarb
- Raw or undercooked beans and legumes
- Raw or undercooked rice
- Citrus fruits
- Raw meat
- Fried, sugary, or other junk food
- Moldy or spoiled food

Starting a home flock

Humans share the Earth with an estimated 30 billion chickens. They are, in fact, the most numerous bird on the planet, and are almost exclusively kept in captivity. It stands to reason that these extremely common domestic birds are now becoming popular family pets.

Driven for many by a desire to connect to their sources of food and gain independence, backyard chicken keeping has surged in popularity in recent years. For the first time in nearly a century, small family flocks are becoming commonplace again. While this resurgence has been in progress for the past decade or so, it reached unprecedented levels in 2020. During the COVID-19 pandemic, so many households began keeping chickens that hatcheries were unable to keep up with demand.

If you're thinking of taking up chicken keeping, you're in for a delightful experience. Chicken hobbyists of all ages and experience levels enjoy numerous benefits, including better physical and mental health, more bountiful gardens, and of course a nutritious source of meat and eggs. Whether you're looking for a sustainable source of food or simply companionship, there are many reasons to consider keeping chickens.

The benefits of chicken keeping

Arguably the most common motivation for establishing a backyard chicken flock is to have access to eggs. It's true that a free-range backyard chicken will lay fresher, more nutritious eggs than what can typically be found in a store—eggs from backyard hens that have access to fresh grass and bugs are up to three times higher in Omega-3 fatty acids and contain less than half the cholesterol of commercial eggs. Many claim that fresh eggs are much tastier, too!

For those keen to support greater sustainability within the home, in their community, and for the environment, keeping a flock of chickens is worth serious consideration. By harvesting meat or gathering eggs from your own flock, you withhold support for commercial factory farms, along with the poor conditions and groundwater pollution that often go with them. You also allow yourself the opportunity to trade, sell, or give away your meat or eggs to your neighbors.

Chickens offer sustainability beyond producing eggs and meat, too. They are ravenous omnivores, capable of eating just about anything, so chicken keepers can drastically reduce the waste their household produces by giving leftovers to their chickens (see page 235). The chickens in turn use the energy from food to lay more nutritious eggs and produce nitrogen-rich compost. This also reduces the amount of feed you need to buy for them. The compost that chickens produce is considered to be one of the best fertilizers available, and a flock of chickens will even help turn the compost and prepare the soil for planting, and will happily gobble up every grasshopper and caterpillar they find. With all these benefits, it's no wonder that chickens and kitchen gardens have existed synergistically for centuries, and are now such an important part of sustainable food systems.

Chickens also contribute to the mental well-being of their keepers. There's something very soothing about watching a flock of hens wandering around a backyard—akin, perhaps, to watching fish swimming in a pond. Their contented clucks and soft feathers are enough to make anyone smile, and the simple act of caring for them can help keep minds and bodies more healthy. Chickens are surprisingly affectionate, too, especially when raised from chicks. They love to come running when their favorite human appears with treats in hand, and are known for cuddling in laps. They therefore make excellent outdoor pets for those simply seeking companionship.

Getting started

Simply getting started is by far the most difficult part of chicken keeping. Chickens require a great deal of investment upfront in terms of time, labor, and finance. Once a safe, healthy environment is established, however, they can be kept without much effort.

The very first step every aspiring chicken keeper must take is to become familiar with the relevant laws and ordinances in their local area. Chickens are generally considered livestock, and many communities still perceive them as unsuited to urban neighborhoods. Thanks to their growing popularity, many cities around the world now accept poultry as

Reasons to keep chickens

Eggs: Eggs from backyard chickens are fresher and more nutritious than store-bought eggs.

Meat: Backyard keepers can ensure their meat birds are raised and harvested humanely.

Gardening: Chickens produce excellent fertilizer and reduce garden pests.

Sustainability: Chickens help reduce food waste and reliance on imported food.

Mental health: Caring for chickens helps lift moods and encourages more time spent outside.

Companionship: Chickens can be cuddly, affectionate, and fun for all ages.

Small home flocks have seen a resurgence, as individuals and families seek to connect to more sustainable food sources, bountiful gardens, and companionship, among myriad reasons.

pets, but local regulations must be taken into account. Most cities prohibit roosters, and have strict regulations on flock size and housing. In addition, local health departments, covenants, and homeowners' associations have their own rules regarding chickens. To prevent costly legal battles and the risk of losing your beloved birds, it's important to know and abide by the rules. If you're unsure where to look, visiting your state or city's official website is a good starting point. Local libraries can also be a good resource.

Once you have a full grasp of neighborhood poultry laws, you'll need to decide what you want most out of chicken keeping. This will help determine flock size, housing requirements, and which breeds to choose. This is also an important time to consider the long-term commitment that keeping chickens requires. An average hen will live eight to ten years, and chickens live outside. That's a long time to have to check on your animals every day, no matter how bad the weather may be. And while a small flock is fairly self-sufficient, they still need fresh water and food daily, along with basic health care and regular coop cleaning. This not only means dedicating time out of your busy schedule when you're at home, it also requires hiring outside help when you're away, and this expense can add up quickly if you travel often. As any chicken enthusiast will tell you, though, they are well worth the cost and effort.

Selecting breeds for your needs

There are an estimated five hundred breeds of chickens around the world today, so it's reasonable to assume there's a chicken breed for everyone. But with so many options, it can be hard to choose. For beginners and families, it's a good idea to consider the more easygoing breeds to begin with. Sticking to these breeds will ensure birds with a gentle, friendly temperament who lay well and don't make too much noise. They generally fall under the heritage and dual-purpose breed categories. These are the same types of chickens that families used to raise on their homesteads for hundreds of years. They include dozens of breeds, but some of the most popular include Orpingtons, Australorps, Rhode Island Reds, Plymouth Barred Rocks, Brahmas, Faverolles, Wyandottes, and Dominiques (all covered in Backyard Chicken Breed Profiles, from pages 48 to 189).

For keepers who want to focus primarily on egg laying, sticking to breeds that lay in excess of 250 eggs per year is the way to go. These are the breeds preferred on large-scale egg farms. The hens are more likely to lay year-round and rarely go broody, ensuring maximum production. Keep in mind, though, that these breeds also tend to be less hardy, more difficult to handle, and generally have shorter life spans. Still, some can be very friendly, and all are exceptionally efficient, producing the most eggs for the least amount of feed, especially if they are allowed to free range. Breeds best known for their laying ability include Leghorns, Production Reds, Gold Comets, and sex links.

A handful of breeds in the egg-laying category deserve special mention here—not because they lay as well as commercial birds, but because they lay uniquely colored eggs. Olive Eggers are a mixed breed that lay dark, olive-colored eggs. Marans lay striking, dark chocolate-colored eggs, while Ameraucanas, Araucanas, and Legbars lay bright green and blue eggs. Mixing one or more of these breeds into your flock will result in a delightfully colorful egg basket.

Chickens have been kept for their showy looks and companionship for thousands of years, and many breeds truly excel as pets. If you're primarily looking for a colorful, fun flock, virtually all bantams (miniature chickens) are ideal, as are many fancy standard birds. There are plenty of breeds that fall into this category, but some of the favorites in the chicken-fancy community include Silkies, Sebrights, Japanese, Polish, Sultans, and Old English Game fowl.

Chicks versus adults

Once all the foundations are covered, it's time to decide whether to start with chicks or grown birds. Both chicks and adults have their pros and cons; it simply comes down to individual preference. As long as the birds have the proper environment, both are excellent options when starting out.

Baby chicks

Is there anything cuter than a tiny chick? Every spring, hatcheries and feed stores around the world show off these balls of fluff, and it's always tempting to go out and buy a batch. Many chicken keepers choose to start out with chicks, and not just for their adorable looks. For one, chicks are much easier to tame and raise into friendly adult birds. Mail-order chicks are also often the only way for keepers to get the specific breeds they're looking for, especially if those breeds are rare. Chicks are significantly cheaper than adult hens as well, usually ranging from US$2 to 6 each (up to roughly £5 in the UK).

Baby chicks are not for everyone, however. They are extremely delicate and require specialized care, with regular check-ins throughout the day, especially for the first six weeks (ideally every few hours for this stage). Unlike adult birds, chicks lack the ability to regulate their body temperature, and therefore need to be housed in brooders with heat lamps or brooder plates (see page 243). They also need to be kept indoors, and not everyone is fond of the smell or the mess. Because they're so delicate, flocks of young chicks are also much more likely to experience illness and fatalities. Lastly, sexing chicks is usually not 100 percent accurate, so you may end up with potentially unwanted roosters.

Pullets and hens

For families looking to avoid the challenges of raising chicks, and who wish to get a head start on those farm-fresh eggs, adult birds are the way to go. Laying hens and pullets (young hens not yet laying) can be put in the coop the day you get them, no brooder setup required. Sexing at this age is guaranteed, too, so you can avoid any unwanted roosters. Most of the time, hens and even pullets will start laying anywhere from a day to just a few weeks after you bring them home, which is significantly faster than waiting for chicks to reach maturity at twenty to forty weeks.

Some keepers do avoid starting out with mature hens, though. For one, laying hens are more expensive, costing US$30 (around £24) or more. Mature birds are also difficult to age, and it isn't uncommon for farms to rehome hens who are past their prime. Getting pullets and young hens from a reputable source is the best way to avoid this issue. Lastly, those who prefer friendly, tame chickens will probably want to avoid adult birds, which are generally difficult to handle and don't bond as well to their families as young chicks.

Housing and location

As a first-time chicken keeper you'll quickly realize that housing is one of the most costly—and most limiting—factors of owning a backyard flock. A well-constructed and well-positioned chicken coop can make the difference between a positive experience and a stressful one. Chickens require a minimum of 2 to 4 square feet (0.2–0.4 square meters) of space per standard-sized bird inside an enclosed coop, with an additional 8 to 12 square feet (0.7–1 square meters) of space per bird outside to roam. Most experts agree, however, that backyard chickens should be given at least twice that amount of space to avoid common issues like bullying and excess noise.

As an example, a comfortable square footage for a backyard flock of five chickens would be:

A 5 x 4-foot (1.5 x 1.2-meter) coop.
A 10 x 10-foot (3 x 3-meter) chicken run.
Plus additional safe space to free range, if possible.

The materials and construction quality of chicken housing matter a great deal as well. While low-cost, prefabricated coops are a possibility, larger coops that are draft-free and predator-proofed will last much longer and are less likely to result in injured or stressed birds. The same rule applies to the chickens' outdoor enclosure. Much consideration needs to be given to free-ranging versus secure enclosures (see pages 218–24).

Lastly, you will need to consider the location of the coop within your backyard. Coops should ideally be located in a shaded area protected from strong winds, where they are close enough for easy access, but far enough away from the house that smell and noise don't become an issue.

Selecting the right chicken age, in addition to the right breed, will help get your flock the best possible start. Keep in mind, as long as your flock is not too large for your space, you can also always add a new breed or age later on. Just be aware of the infamous "chicken math" problem that keepers around the world struggle with. Somehow, it makes you want to get more chickens than planned!

Above While startup costs for chicken-keeping equipment can be expensive, it can last for many years. **Below** Adequate brooding equipment is essential for chicks' survival for the first eight to twelve weeks.

Chicken-keeping equipment

Getting started with chickens involves a significant upfront investment in chicken-keeping equipment. A coop and wired enclosure in particular can be prohibitively costly, depending on the size and design. If you start your flock with young chicks, brooding will require additional equipment and expense. Fortunately, once everything is in hand, you'll have all you need for a healthy flock, and it will last for many years, over many generations of chickens.

Brooding chicks

Chicks are much too delicate to be housed outside, unless under the protection of a mother hen. For the first eight to twelve weeks of their lives, they need to be kept indoors, safe and warm in a brooder, and supplied with the appropriate nourishment.

Brooders

A secure brooder is extremely important for the health and safety of young chicks. This is where they'll spend the majority of their days until they're old enough to go outside. Store-bought brooders are widely available; however, most chicken keepers prefer to make their own. Homemade brooders can be constructed out of wood, or made from a repurposed large plastic tote, dog crate, or even a pop-up tent. Hard-sided brooders are generally preferred for the first two or three weeks of a chick's life, when they are most susceptible to drafts and chills. Then, as they get bigger and more active, large open-sided enclosures can give them the space they need. All brooders should be secured with a wired top to prevent the chicks from flying out, and must be kept safely away from children and other pets.

Heat sources

Like all baby birds, chicks are not able to regulate their own body temperature and can die quickly if not kept warm enough. They require a safe space where the temperature is kept at a steady 90 to 95° Fahrenheit (32–35° Celsius) for the first week after hatching. This heat is mandatory and must remain constant. Gradually reduce the temperature by 5°F (just below 3°C) degrees per week until it reaches

70°F (21°C), after which the birds will be fully feathered and able to regulate their own body heat. The heat source itself can be either a heat lamp or a plate brooder designed for chicks. Both are good options, provided they are safely set up and secured. Inspect the heat source daily for any signs of dust buildup or damage.

Thermometer

To ensure that the temperature remains at adequate levels throughout the brooding period, a thermometer placed in the brooder is highly recommended.

Bedding

Chicks require a substrate at the bottom of their brooder to help absorb droppings and provide insulation against a cold floor. Bedding materials can vary from straw and aspen shavings to paper towels and shredded newspaper. Avoid using cedar shavings, which are toxic to chicks. Sawdust and sand are also discouraged, as they can generate harmful dust and cause crop problems should the chicks start to eat them.

Food and food containers

All baby chicks should be fed a chick starter and grower (see page 233). This feed formula provides balanced nutrition and is gentle on their sensitive digestive systems. Chicks should never be offered layer feed—its high calcium content will cause liver problems. Food should be offered in a shallow container, ideally one designed for chickens so as to reduce spillage and waste. Make sure that food is clean and available at all times.

Water and water containers

Chicks also require constant access to fresh, clean water. They are very prone to drowning, however, especially when very young, so to prevent accidents, use a shallow waterer designed for chicks, and keep it secured so it cannot tip over (see page 232). Keep in mind, chicks are very messy and the water will likely need to be replaced throughout the day.

Dust bath and enrichment

Chicks are inquisitive creatures and display adult-like behaviors very early on. Keeping them happily occupied will prevent behavioral issues and prepare them for a life outdoors. After the first week or so, for example, chicks benefit greatly from a dust-bathing area in their brooder. Offer them a shallow dish of fine dirt from outside, and watch as they enthusiastically begin digging and bathing. Offering short roosting bars and sticks can also help them practice grown-chicken behaviors, like perching and balancing.

Adult chickens

Once fully feathered, chickens should be housed outside with adequate food, water, and shelter. Setting your flock up for success with the proper equipment right from the start will result in longer, healthier lives, and far less stress from issues like predation, frostbite, and illness.

The coop

No chicken flock is safe without a solidly built coop. This is a fully enclosed structure where they sleep at night, lay eggs, and escape the elements. It is, in essence, a small barn designed for chickens. Quality chicken coops can be purchased from a store, custom built (see page 275), or established in a repurposed shed. They come in a wide range of sizes and designs, but they all have the following in common:

Solid construction

To keep your chickens warm, dry, and safe from predators, the coop must have solid walls and a roof. There should be no drafts, and no way for moisture to seep in. Ventilation, which allows air flow up high near the ceiling, should be provided, but secured with wire to prevent predators and pests from entering. It's possible to have a coop built with just the ground for a floor, especially in warmer climates, but this is strongly discouraged. A solid floor is much safer and more sanitary. The coop must be accessible at all times for the chickens, with a chicken-sized door near the coop floor, and it should also be easy for you to access and clean. Size matters, too: there must be a minimum of

2 square feet (0.2 square meters) per chicken to prevent overcrowding.

Bedding

For optimal comfort and cleanliness, the coop floor should be lined with bedding. As with the chick brooder, there are many options available. The most popular choices are sand, straw, and pine shavings. Sand is arguably the easiest to clean—using a kitty-litter scoop—but pine shavings and straw provide better insulation and are compostable as well. No matter which bedding you choose, it must be kept reasonably clean and clear of moisture, mildew, and pests.

Roosting bars

Chickens naturally "roost" at night, perched up off the ground and away from predators, so domestic chickens need to be given this opportunity with the addition of roosting bars inside the coop. These can be either thick dowels, tree branches, or lumber (4 x 2-inch planks are a good size). Larger breeds will need roosting bars that are thick and strong enough to support their weight comfortably.

Nesting boxes

Every coop housing hens needs a safe, comfortable place for them to lay their eggs. The construction and location of any nesting boxes isn't particularly important, as long as they're easily accessible by both hens and their keepers. Ideally there should be one nesting box per four to five hens, though you may find that some boxes are preferred over others. The boxes must contain some bedding or a soft floor to prevent breakage and they should be cleaned regularly.

Laying lights and heaters

Laying lights and heaters are not necessary (and often controversial), but they are possible additions to a chicken coop. Laying lights give off soft light inside the coop for several hours at night during the winter, encouraging hens to lay year-round. This is a well-accepted practice, especially on hobby farms and egg farms where a year-round supply of eggs is needed. Some backyard chicken keepers advise against this

A well-constructed chicken coop is essential for chickens to stay safe and warm, as well as for laying eggs. In addition, room to roam (enclosed or free range) and sufficient fresh water and food are key for a happy, healthy flock.

practice, however, as it may stress hens at a time when most of their energy should go toward keeping warm.

Heaters are also widely accepted, but they can be dangerous if not used properly (see page 229). If your situation warrants using a heater in the coop, avoid using standard heat lamps, which are a serious fire hazard in outdoor coops, and purchase a lower-wattage panel heater instead. These are designed for chicken coops and are significantly safer. Other sources of heat, such as heated perches, are known to cause burns or fires and are best avoided.

The chicken pen

An enclosed outdoor pen (also known as the "chicken run") for chickens is not required (see page 223), but for many keepers it's a necessity. The outdoor pen is where all food and water should be kept, and where your chickens will spend the majority of their day. Unless the chickens free range most of the time, the minimum size requirement for an enclosed pen is 8 to 10 square feet (0.7–0.9 square meters) per chicken—the bigger, the better.

Food containers

New chicken keepers will quickly find that there are as many options for feed containers as there are brands of feed. There is no one feeder that's best for every situation. Many keepers prefer large gravity feeders, such as those purchased from a store or made using PVC pipe. Others prefer to use no-waste feeder systems that open up only when a chicken steps on a pressure plate. Still others are happy with a simple horizontal feeder. As long as food is readily available and easily accessible, your chickens will do just fine.

Water containers

Clean water must be provided at all times. Because each chicken can drink around a pint a day (just over half a liter), a large gravity waterer is highly recommended. For places where temperatures are routinely below freezing in winter, a water heater may also be necessary.

Outdoor litter

The chicken run should be constructed directly on the ground, allowing chickens easy access to the dirt for scratching and foraging. Any vegetation in the run will be destroyed in short order, and the bare ground will get messy quickly, so most keepers prefer to lay down substrate. The most popular material for this is sand, followed by wood, mulch, and straw. Pine shavings break down quickly and are generally impractical for an outdoor pen. Whatever litter type you choose, it should be refreshed and replenished at least once every few months, depending on the size of the run and whether it's exposed to rain and snow.

The deep litter method

For chickens residing in colder climates, a chicken waste-management system known as the "deep litter method" can be an effective strategy. The idea is to take advantage of the natural heat generated by composting to keep the coop warm. First, add a thick layer of bedding, starting in fall. Then turn the soiled bedding at least once weekly and add a layer of fresh, clean bedding on top. Over winter the droppings and bedding will decompose and generate a bit of heat. This system is both easy and economical to maintain, and it creates a rich compost for the garden come spring.

Grit and calcium

Nutritional supplements such as grit and calcium (see page 234) should be made readily available. These don't need a special container—even a small tupperware box is fine—as long as your chickens can access it easily.

Dust-bathing area

All chickens must have access to soft dirt and sand to take frequent dirt baths (see page 204). Most keepers prefer to construct a dust-bathing area using a shallow box, planter, or even an old tire. The dust bath should be large enough to accommodate a few chickens at once, and must be kept protected from rain and snow. Store-bought bags of dust-bathing material are available, but any mixture of topsoil and sand is perfectly fine. Make sure to replenish it as needed.

Enrichment

Chickens are very active, inquisitive creatures, and benefit greatly from fun activities in their run. Try incorporating some large branches to provide outdoor roosting opportunities and to capitalize on vertical space. Smaller, flightier breeds such as most bantams are especially fond of perching in high places. You can also incorporate special treats, such as "flock blocks" (solid blocks of grain), which are available at most feed stores and will reduce both boredom and excess noise.

First aid kit

Health problems are not a certainty in a small chicken flock, but they're definitely always a possibility. A 2010 study by the US Department of Agriculture found that about 25 percent of urban flocks experienced health problems. Fortunately, the issues faced by smaller backyard flocks tend to be both fewer and easier to handle than those seen on large-scale farms, and they can usually be treated with a simple chicken first aid kit. You can either purchase a kit or put one together yourself, as long as it includes the following:

- Small gauze pads
- Vet tape
- Syringe
- Epsom salt
- Antiseptic ointment
- Styptic powder or cornstarch (cornflour)
- Electrolytes
- Wound care spray
- Small nail clippers
- Scissors
- Small crate or carrier

CHAPTER THREE

CHICKEN-FRIENDLY PROJECTS

THE CHICKEN-FRIENDLY BACKYARD

The quintessential sight of happy hens wandering around a cozy cottage garden is the sort of soothing experience that every backyard chicken keeper yearns for, but this vision can unravel quickly if precautions are not taken—chickens and backyards don't always get along seamlessly. However, with a bit of planning, that dreamy backyard chicken experience can become a reality.

Selecting a location

A chicken-friendly backyard begins with a good location for the chicken coop and run. A well-positioned home for your chickens can make the difference between a happy and comfortable chicken-keeping experience and a stressful one. Coops are

Opposite and above With proper planning and the right equipment and care, a chicken-friendly backyard can become a reality.

generally very heavy and difficult to move, so select your location carefully. Several factors must be taken into account when deciding on the location of the coop and the run, starting with local laws and ordinances (see page 19), and the weather conditions affecting your backyard.

In many cities, a coop must be positioned a minimum distance from neighboring houses and fence lines. Chicken coops can also be smelly, even when well maintained, so consider building yours some ways away from your kitchen and bedroom windows. Many keepers opt to build theirs far from the house; however, you'll need to consider convenience as you'll be visiting your coop at least once every day.

Your birds' comfort should be taken into account when considering location, too. An ideal position will be in the shade, such as under a tree, or at least in a generally protected area where wind and exposure to the elements can be minimized. If no such location exists in your yard, you'll need to compensate by providing additional protection with, for example, a roof, sunshade, or wall panels on the run. Lastly, excessive mud can cause problems for both chickens and their keepers, so make sure you take drainage into account.

Keeping chickens and gardens safe

Safety is a two-way street when it comes to keeping backyard chickens. These birds can be quite destructive and pose a serious threat to prized lawns and plants. Conversely, well-maintained gardens can pose several safety hazards to chickens. The key to avoiding stress—or tragedy—is to plan accordingly, leveraging barriers when needed, and setting up a system where both chickens and gardens can benefit one another.

Protecting your garden

Expert keepers know that nothing destroys a garden faster than a flock of hungry foraging chickens. Within days, these birds can significantly impact the landscape. Their strong, sharp claws will fling mulch and gravel in all directions in a quest to find bugs, and they'll dig large holes in flower pots and garden beds to dust-bathe, uprooting any plant in the way. Being ever-hungry omnivores, chickens are constantly on the lookout for tender, tasty leaves, fruits, and seeds, and they'll happily use their sharp beaks to defoliate precious shrubs, devour flowers and fruits, and graze a manicured lawn down to a patchy mess.

If you value your garden's appearance, protection from marauding chickens is key. Fortunately, these birds, while persistent, are generally dissuaded by chicken wire, mesh, and similar barriers. They can also be discouraged from flying over low fences—around 4 feet (1.2 meters)—by leaving the tops flimsy so that the birds cannot perch on them, though for flightier breeds this might not

work. A short picket fence topped with chicken wire, for example, can be enough of a deterrent to prevent chickens from accessing a vegetable garden.

When it comes to lawns and large garden beds that aren't as easy to protect, the best course of action is to limit the birds' access to them—either by restricting the amount of time they free range each day, or by keeping them confined in a moveable run (see page 223).

Protecting your chickens

Aside from predators (see page 218), various potential hazards exist for chickens in a typical backyard. As with other pets and children, poisonous plants are a common concern. Chickens are generally savvy and avoid toxic plants, but mistakes can happen and accidental poisoning may occur if these plants are accessible to your birds (see box opposite).

As with poisonous plants, many of the other common hazards for chickens in a typical backyard are best removed completely to ensure their safety. Common examples include mothballs, mouse poison, and the granular pesticides included in lawn revitalizers. These are all very toxic, and chickens will eat them if given the opportunity. Other pest-control products can be hazardous as well, including mouse traps and sticky insect traps, both of which can lead to serious injury. In general, products labeled child- or pet-safe are a better option, but it's still important to use good judgment and keep all potential hazards out of reach.

Keeping your backyard clean

Even with every safety measure taken, chickens are messy creatures, and they make their presence known in every backyard they inhabit. If they are allowed to roam free, additional weekly chores will be needed to keep your backyard clean.

Plants to avoid

Common backyard plants to keep away from chickens:

Nightshades
Foxgloves
Lupins
Pits from stone fruits such as cherries
Azaleas
Beans and other legumes
Oleander
Daffodils
Bracken fern
Lobelia
Holly
Periwinkle
Rhubarb
Yew

Top left Rhubarb **Bottom left** Foxgloves **Bottom right** Daffodils

Droppings

The most obvious mess that chickens create is with their droppings. An adult chicken defecates once every twenty minutes on average. This equates to a lot of droppings throughout the yard every day, even for a small flock. Unfortunately, chickens cannot be "potty trained" to defecate in one area only; they will leave their droppings virtually everywhere they go, and trampolines, childrens' play structures, back porches, and outdoor tables and chairs are frequent targets.

Dried droppings on lawns or walkways are usually easy to remove using a rake or broom, and for everything else a handheld scraper and stiff brush generally does the trick. A vinegar solution or natural multisurface cleaning spray is also effective for removing droppings from hard outdoor surfaces. To avoid having to clean up daily, many keepers opt to protect their favorite areas by preventing their chickens from leaving a mess in the first place. Chairs, for example, can be stored upside down on an outdoor table when not in use, and safety mesh can be installed as a barrier on trampolines and other play structures.

Feathers

Chickens will occasionally drop feathers throughout the year; however, all adult chickens will shed an excessive number of feathers twice a year in a seasonal process called "molting." This usually occurs in spring and fall and lasts for several weeks, during which time the coop and backyard will be overrun with hundreds of feathers, getting stuck in lawns and shrubs, and gathering in piles everywhere. Like with fall leaves, feathers are relatively easy to remove with a wide rake; just make sure to clear them on a windless day when they are less likely to blow around.

Chickens and gardens existing in harmony

Chickens and gardens may present hazards to one another, but when well managed, a chicken-friendly backyard can enhance the health of both chickens and plants, leading to bigger and more delicious harvests from your garden.

Composting

One of the greatest side benefits of keeping chickens is that their droppings can be used as manure; being high in nitrogen, they make an exceptional and sustainable fertilizer. Feathers and eggshells are also excellent additions for your garden compost, adding calcium and other nutrients to promote healthy plant growth. Simply add these to your compost pile every week after cleaning. If you use the deep litter method (see page 246), composting can be done directly in the coop or run, with the chickens doing all the hard work! After a few months, simply remove the compost with a shovel and add it to the garden. It's important to note that chicken fertilizer is considered "hot" when fresh. This means that the nitrogen content is too high initially and it will cause damage to plants—commonly known as "fertilizer burn"—if applied too early. Therefore, you should always compost chicken manure properly before adding it to your garden.

Pest control

Another way that chickens can benefit a garden is through pest control— particularly large pests such as grasshoppers, grubs, and caterpillars, which are among their favorite foods. Chickens are experts at uncovering grubs in the soil, and they make quick work of a grasshopper infestation. To remove soil-dwelling pests before planting, simply open the garden bed to your flock of chickens and let them dig up hidden pests under the topsoil. During the

growing season, chickens are generally too destructive to hunt pests in garden beds unattended, but providing some limited access can be very helpful. Alternatively, you can capture garden pests, then feed them to your chickens as a treat. This is an excellent way of disposing of hornworms (the caterpillars of hawk moths) and beetles.

Once the growing season is over, your chickens can be given access to the garden again, allowing them to remove the last of the season's garden pests before they lay eggs for the following year. If you use cover crops during the winter to improve the soil, keeping chickens can double your benefits. Once the cover crops are finished, let the chickens graze on them. They will provide a valuable nutrition boost when other fresh foods are harder to come by. Just make sure the cover crops you use are safe for chickens to eat. Alfalfa and clover are excellent choices. (See page 266, for more on growing plants for your chickens' consumption.)

Above If properly managed, a chicken-friendly backyard can be advantageous for both chickens and plants, by removing pests and supplying fertilizer, while providing additional sources of recreation and sustenance for the chickens.

Planting for chickens

As mentioned, there are lots of good reasons for gardening with chickens, from improving the aesthetic appearance of chicken runs to creating the ideal natural habitat for your flock to shelter, roost, forage, and socialize in. In return, chickens can help maintain a healthy ecosystem by snacking on weeds and bugs, help rotavate the soil while pecking around for food, provide nutrient-rich manure courtesy of their droppings, and, of course, reward with eggs.

On the flipside, chickens can wreak havoc on the best-laid garden plans, pulling up saplings, decimating new shoots or leaves, or causing chaos in the vegetable patch or cut flower bed by eating everything in sight. Some plants or plant parts are also toxic to chickens and thus need removing from your plot or keeping well out of reach.

Gardening with chickens can therefore be very much a case of trial and error, experimenting with species that are safe, productive, and robust enough to survive. It's also worth thinking about your chickens' rainforest roots (see Bamboo, below), to understand more about their inherent urges and desires: the need to forage and explore to fulfil an omnivore diet, find safe and shaded places in which to make dust baths and nests, and find perches on which to roost and preen.

Trees, hedges, and shrubs

Many of your chickens' needs can be satisfied by the inclusion of trees and hedges. In larger gardens this could include full-size fruit trees or a multipurpose screen of bushes or hedging plants to create shelter belts that will help cut down wind chill, attract beneficial wildlife and insects, deter predators, and reduce erosion. In smaller spaces, there will still be plenty of opportunities to grow small trees or shrubs such as bamboo or butterfly bushes in containers or pots. The benefit of this is that specimens can also be moved around if you need or want to adapt your plan. Climbing or rambling plants such as roses or grapevines are also great for growing up structures such as hen houses or coops, providing delicious edible or medicinal treats for you and your chickens.

Opposite Give careful consideration to the plants in any chicken-friendly garden. Fortunately, there is a wealth of choice in the trees and shrubs, flowers, grasses, and edibles that are suitable for a home flock.

Ten to try

1. Mulberry tree
This deciduous, heart-leaved, fruiting tree, native to Mesopotamia and Persia, produces compound clusters of black, red, or white fruits that are loved by chickens as well as humans. Let your birds forage under the branches of a mature specimen to eat their fill of these tasty, antioxidant-rich fruits, or portion them out as treats. Choose the black mulberry (*Morus nigra*)—the red mulberry (*Morus rubra*) and white mulberry (*Morus alba*) produce mildly toxic sap. And be prepared for purple droppings!

Suggested species
Black mulberry (*Morus nigra*, including "Chelsea" or the naturally dwarfing "Charlotte Russe")

2. Siberian pea shrub
Growing to a height of around 10 to 15 feet (3 to 4.5 meters), with oval leaflets and yellow snapdragon-shaped blooms, the cold-winter-hardy Siberian pea shrub or pea tree is ideal for planting as a windbreak, in borders, or to make a flowering hedge or screen near your coop. There are also weeping varieties such as "Pendula" that can be worked to form a small tree. Chickens enjoy the peas, rich in protein and fatty oils, which they can gobble up directly from the ground or be given as a snack.

Suggested species
Siberian pea shrub (*Caragana arborescens*, including the weeping "Pendula")

3. Bamboo
Before planting bamboo, check local guidelines in case of restrictions. This may require planting bamboo in pots. Mimic the natural habitat of early domesticated chickens—widely agreed to have evolved from species of wild junglefowl native to the rainforests of China, Southeast Asia, and South Asia—by planting bamboo. Choose clump-forming tropical or temperate woody bamboos such as *Fargesia*, short-form *Shibataea*, or medium

Clockwise (from top left) The leaves of a robust beech tree, the autumnal leaves of a filbert hazel tree, springtime blossoms on an apple tree, and the red berries of a Washington hawthorn tree are all safe and hardy for a chicken-friendly garden.

to large *Bambusa*. Not only do these bushy plants provide shade and shelter, the leaves—high in calcium, potassium, iron, and zinc—help keep bones and beaks healthy.

Suggested species
Umbrella or fountain clumping bamboo (*Fargesia* spp.); short clumping bamboo (*Shibataea* spp.); large clumping bamboo (*Bambusa* spp.); blue bamboo (*Himalayacalamus hookeriana*)

4. Beech
Chickens love scratching around under hedges foraging for insects and making dust baths. Beech hedges provide excellent cover in or around chicken runs, although young plants will need protecting with tubing or chicken wire to prevent them from being nibbled or pulled up while establishing. The leaves are safe to eat if your birds do find a way to feast on them, and the three-angled beechnuts (also known as beech mast) are safe to eat in small amounts. A hornbeam hedge is another option.

Suggested species
Beech (*Fagus sylvatica*); hornbeam (*Carpinus betulus*)

5. Hazel
Hazel hedging is another good option for creating screening, shade, and dust-bath locations, with the bonus of providing a feast of nutritious nuts for your chickens if squirrels and other birds don't get to them first. If you're growing nuts to harvest, running chickens under the trees in fall can also help disturb nut weevil pupae, thus protecting crops from invasion. Chickens will scratch around at the base of new plants, so ensure they're adequately protected until mature.

Suggested species
Common or European hazelnut (*Corylus avellana*); filbert (*Corylus maxima*); American hazelnut (*Corylus americana*); beaked hazel (*Corylus cornuta*)

6. Hawthorn
Lobed-leaved hawthorn is another deciduous tree that can be grown to maturity or planted up as a hedge, on its own or with other native species. The pink-white spring flowers attract lots of insects, while the red haws that develop in fall can be safely eaten by chickens foraging underneath, or given to them fresh or dried as an antioxidant-rich treat. As above, protect new saplings from vigorous pecking using tubing or chicken wire.

Suggested species
English hawthorn (*Crataegus laevigata*); common hawthorn (*Crataegus monogyna*); Washington hawthorn (*Crataegus phaenopyrum*), downy hawthorn (*Crataegus mollis*); cockspur hawthorn (*Crataegus crus-galli*)

7. Rose
What could be prettier than a chicken coop or run covered in climbing or rambling roses, or ornamented with tubs of fragrantly flowered rose shrubs? This is the symbiotic power of nature at its best, where the roses benefit from a ready supply of nitrogen-rich chicken manure and the chickens get to snack on highly nutritious rose hips, aromatic rose petals, and any visiting bugs. Ensure young plants have a chance to establish, protected from pecking beaks and scratching.

Suggested species
Good hip-producing species include Japanese rose (*Rosa rugosa*); dog rose (*Rosa canina*); *Rosa filipes* "Kiftsgate"; *Rosa rugosa* "Fru Dagmar Hastrup"; Moyes rose (*Rosa moyesii*)

8. Butterfly bush
If you're looking for a flowering shrub for the chicken run, coop, or poultry pasture that won't get decimated by hungry, curious, or bored chickens, the butterfly bush could be it. Its purple, pink, mauve, or white cone-shaped flowers are pretty and attract tasty bugs and butterflies, while the hardy trunks and stems can withstand a bit of pecking, and they are fast-growing, so ideal if you want to create shade and shelter quickly. There are also a few different species to choose from.

Suggested species
Butterfly bush (*Buddleja davidii* including "Dart's Purple Rain," "Pink Delight," and "Royal Red"); Chinese buddleja (*Buddleja albiflora*; *Buddleja delavayi*); alternate-leaved butterfly bush (*Buddleja alternifolia*)

9. Grapevine

Unlike other pets such as cats and dogs, chickens can eat and actively benefit from antioxidant-rich grapes, making this beautiful and useful climber another good option for ornamenting your chicken run or coop. Protect saplings and young shoots and leaves from pecking and scratching, and avoid too much sugar intake by training the fruiting sections of your grapevine arbor away from accessible posts or high perches. Never feed whole grapes to chicks as they can choke on them.

Suggested species
Common grapevine (*Vitis vinifera*); frost grape (*Vitis vulpina*); muscadine (*Vitis rotundifolia*); California wild grape (*Vitis californica*)

10. Apple tree

Investing in an apple or crab apple tree brings multiple rewards in the form of pretty spring and summer blossom, attractive bark and leaves, and an annual harvest of delicious fruit. Feed your chickens picked apples as a treat (chopped up minus the core and seeds, or hung whole) or consider letting your flock run free range under the trees at harvest time to help make good of any windfalls, eat residual bugs, and help naturally fertilize the ground with their droppings.

Suggested species
Edible apple (*Malus domestica*, including numerous cultivars); European crab apple (*Malus sylvestris*); southern crab apple (*Malus angustifolia*)

Flowers, grasses, and edibles

Although creating a chicken-friendly garden can be a challenge, take your lead from nature and you can have the beautiful backyard you've been dreaming of—for you and them.

Start with getting to know your chickens. What do they like to eat (see page 232)? Think about the kind of foods that are present in store-bought chicken feed, such as grains and seeds, plus the supplementary snacks and treats you provide, such as leafy greens, beans, corn, berries, and apples. Could you grow any of these items yourself?

Observe your birds' habits in the run or when left to free range. Chickens love to forage for edible plants and grubs, scratching around for hours as they look for sample suitable morsels. Their dream garden is primarily a forest environment with a canopy of high and mid-range plants, pockets of dense vegetation, low-growing understory plants, and scrubby or shrubby areas where field and forest meet. Ideally, they can move among these different habitats, through the seasons, to find shelter and harvest favorite foods.

Taking a forest gardening approach is one way to recreate this scene, establishing a diverse, multilayered, and sustainable ecosystem of trees, shrubs, and perennial plants, including rotations of edible crops and companion plants. For smaller plots, or keepers who prefer a more ornamental garden, there are still lots of plants you can try, from clover-filled poultry pasture and ornamental grasses to beds or containers of beautiful and beneficial herbs, as well as edible flowers or delicious fruits and vegetables. Do go organic where possible.

It's worth noting that there really is no such thing as a chicken-proof plant—they will always peck at and pull things out. It will take a bit of trial and error before you get the balance right, so be amenable to moving things around or losing the odd shrub or even a whole harvest of fruit along the way. Use common sense when positioning.

While the plants that follow are largely edible or non-toxic, there are also some species that might harm your birds—including members of the nightshade family, such as potatoes, tomatoes, and eggplants, plus onions, avocados, citrus fruits, foxgloves, yew, ferns, ground ivy, and tulips—so always do your research first.

Ten to try

1. Grasses

The ideal poultry pasture for a chicken run or dedicated foraging area is a mix of perennial rye grass, fescue, and clover, providing nutritious

Clockwise (from top left) Start with getting to know what your chickens like to eat and how they like to forage. Consider edible fruit and vegetables as well as flowers and leaves, but be prepared for a case of trial and error to get it right.

Clockwise (from top left) A wide range of grasses, flowers, and edibles make for both a stunning garden and a grazing buffet for chickens. Why not try sunflowers (such as helianthus), lavender (such as French), or stunning white or red clover.

greens and flowers and the insects and grubs harboring within. For more rural or wilder plots, native grasses such as tall fescue or Kentucky bluegrass are ideal (research what is best for your location). For ornamenting and a seasonal harvest of foraged seeds, try borders or pots of oat grass or switchgrass.

Suggested species
Tall fescue (*Festuca arundinacea*); blue fescue (*Festuca glauca*); Kentucky bluegrass (*Poa pratensis*); oat grass (*Stipa gigantea*); switchgrass (*Panicum virgatum*); little bluestem (*Schizachyrium scoparium*); Canada wild rye (*Elymus canadensis*); Indian grass (*Sorghastrum nutans*); tufted hairgrass (*Deschampsia cespitosa*)

2. Ground cover and weeds
Nutrient-dense clover as part of a free-ranging diet is ideal for chickens as they can eat the greens and flowers and any attracted bugs. They also love a range of weeds, including dandelions, broadleaf plantain, and chickweed, which can be picked and added to the coop. Letting chickens have the run of organically grown crop-cover plots can also be beneficial as their foraging habits and nitrogen-rich droppings can help improve the quality of the soil.

Suggested species
Crop cover: White clover (*Trifolium repens*); red clover (*Trifolium pratense*); alfalfa (*Medicago sativa*); annual ryegrass (*Lolium multiflorum*); white mustard (*Sinapis alba*)

Weeds: dandelion (*Taraxacum* spp.); broadleaf plantain (*Plantago major*); nettles (*Urtica dioica*); purslane (*Portulaca oleracea*); wood sorrel (*Oxalis acetosella*); tick trefoil (*Desmodium canadense*); chickweed (*Stellaria media*)

3. Herbs
Growing herbs in or around your chicken run not only makes it look pretty and smell good—some of these plants also have beneficial dietary, antibacterial, antiseptic, and aromatherapeutic properties that can help provide vital nutrients, boost immunity, ward off diseases and pests, and reduce stress. Oregano and rosemary, for example, may help keep respiratory infections at bay, while mint and lavender

hung inside the coop can help repel pests and neutralize odors. The following herbs are all safe to use, with the woody, bushy specimens like lavender or rosemary also providing shelter.

Suggested species
English lavender (*Lavandula angustifolia*); French lavender (*Lavandula stoechas*); rosemary (*Salvia rosmarinus*); common sage (*Salvia officinalis*); Mexican bush sage (*Salvia leucantha*); pineapple sage (*Salvia elegans*); oregano (*Origanum vulgare*); basil (*Ocimum basilicum*); parsley (*Petroselinum crispum*); dill (*Anethum graveolens*); fennel (*Foeniculum vulgare*); wormwood (*Artemisia absinthium*); feverfew (*Tanacetum parthenium*); mint (*Mentha* spp.) thyme (*Thymus* spp.); bee balm (*Monarda* spp.); lemon balm (*Melissa officinalis*); catmint (*Nepeta* spp.); marjoram (*Origanum majorana*)

4. Edible flowers
Like herbs, many edible flowers are safe to have around chickens, helping to beautify their space and provide nutritious and therapeutic treats. Marigolds act as natural garden pesticides, are antioxidant, and can help produce wonderful golden-yellow egg yolks. Adding naturally antifungal and antiviral petals to nest boxes can also help repel pests such as mites or mosquitos and boost the respiratory system. Nasturtiums are a great choice for the coop as they can withstand foraging, and are also thought to enhance the color of egg yolks.

Suggested species
Pot marigold (*Calendula officinalis*); French marigold (*Tagetes patula*); African marigold (*Tagetes erecta*); nasturtium (*Tropaeolum majus*); violets and violas (*Viola* spp.); coneflower (*Echinacea* spp.)

5. Sunflowers
Sunflowers are one of the easiest plants grow (so ideal for growing with children), providing joyful summer color and high-fibre, vitamin-rich snacks for humans and chickens alike. Select black oil sunflowers, which contain the most beneficial fatty acids, or choose giant sunflowers for an abundant crop. Let your chickens peck directly at broken-off flower heads, or dry the flowers first and harvest

the seeds to hand out in moderated portions. Don't worry about shelling them—chickens are able to peck away the tough outer layer.

Suggested species
Black oil sunflower (*Helianthus annuus*); giant sunflower (*Helianthus giganteus*)

6. Amaranth
Amaranth seeds are a good source of protein, vitamins, and minerals and were a staple food for the Aztecs, ground into flour and cooked. The seeds can be fed to chickens in their raw form, and the plant looks fantastic in the garden, with its dramatic tassel-like flowers in shades of crimson, purple, or green, which can be used as cut flowers or left to mature into seed heads. If using as an edible, ensure that plants or seeds are organically grown or sourced. Amaranth is also a good companion crop for corn or maize (*Zea mays* subsp. *mays*), another beneficial, energy-boosting chicken feed.

Suggested species
Amaranth (*Amaranthus caudatus; A. cruentus; A. tricolor*)

7. Leafy greens
Chickens love tender leafy greens such as chard, kale, spinach, and beetroot tops, but if you give them automatic access to your vegetable patch, they're likely to eat them all at once. Instead, plant some away from the coop and allow them to grow to maturity before portioning out handfuls for your chicks to graze on, or harvest them in small bunches and hang them in the run so that they can graze on them throughout the day. A regular treat of dark leafy greens can help produce darker, richer yolks in layers, but it's also a good idea to limit amounts to avoid the buildup of substances such as oxalic acid, which can bind to calcium and lead to a deficiency.

Suggested species
Beet (beetroot) (*Beta vulgaris*); Swiss chard (*Beta vulgaris* subsp. *vulgaris*); kale (*Brassica oleracea*); spinach (*Spinacia oleracea*); radish (*Raphanus raphanistrum* subsp. *sativus*); coneflower (*Echinacea* spp.)

8. Berries and currants
Chickens love snacking on small antioxidant-rich fruits such as raspberries, strawberries, and blueberries. Grow your own crops and you get a source for free, planting away from reach if you want to ration your flock's intake—chickens are capable of gorging on a whole harvest at once, which is not good for them or the plant. Blue and purple berries and currants will color your birds' droppings, so be prepared to deal with the fallout. Low-growing bushes also provide welcome shade while foraging.

Suggested species
Raspberry (*Rubus* spp.); strawberry (*Fragaria* spp.); blueberry (*Vaccinium* spp.); blackcurrant (*Ribes nigrum*); redcurrant (*Ribes rubrum*)

9. Gourds
Keep chickens well fed through summer, fall, and winter with a ready supply of homegrown gourds such as pumpkins, squashes, zucchini, cucumbers, and melons. All members of the Curcurbitaceae family, these diverse plants can be grown in beds or containers, ideally out of the coop so the plants have a chance to grow and fruit. Pumpkin is particularly nutritious, high in beta-carotene and vitamin E via the flesh and seeds, of which chickens can also eat the shell. A snack of orange courgette flowers also boosts yoke color.

Suggested species
Pumpkins and squashes (*Cucurbita* spp.); zucchini (*Cucurbita pepo*); melon (*Cucumis* spp.); cucumber (*Cucumis sativus*)

10. Garlic
Garlic has been fed to chickens for hundreds of years to help promote growth, ward off infection from mites or worms, boost immunity, and counteract respiratory problems. Growing your own crop of garlic in a raised bed or container is a great way to access a regular, organic source. Simply add a crushed clove to your chickens' water a few times a month, acclimatizing them to the strong taste by starting when they're young.

Suggested species
Garlic (*Allium sativum*)

Clockwise (from top left) Where possible, a wide variety of grasses, herbs, fruits, and vegetables can contribute to a richly varied diet for chickens. Consider planting mulberries, amaranth, and gourds, all of which are high in vitamins and minerals.

Make a basic chicken coop

A chicken coop can be as elaborate or simple as your aesthetic and budget allows. The basic plan outlined below provides the perfect-sized coop for a small backyard flock of up to six standard-sized chickens (or up to nine bantams) but the suggested measurements and materials can be adjusted or customized to fit your needs.

Dimensions
- Footprint: 4 x 4 ft (1.2 x 1.2 m)
- Height: 6 ft (1.8 m) at the roof's highest point at the front of the coop, slanting down to 4 ft (1.2 m) at its lowest point at the back

Features
- Chicken–access door
- Two ventilation openings
- 12-in- (30.5-cm-) high open area underneath the coop to keep it off the ground (to help protect chickens from predators and pests)
- Three internal nesting boxes, accessed through a hinged egg-access door
- Swinging side clean-out access door
- Roosting bars

Materials

Lumber (timber): use untreated, non-cedar wood for any part of the coop interior, such as the walls and floor (cedar and treated wood can cause respiratory distress for chickens).

- Four 4x4 posts, 8 ft (2.4 m) long
- Four 4 x 8-ft (1.2 x 2.4-m) panels: siding, plywood or OSB (all preferable to MDF, which decays much too quickly to be used for a coop)
- Twelve to fifteen 2x4s (or 4x2s)—more may be needed for customization
- Two or three 2 x 8-ft (0.6 x 2.4-m) roof panels of chosen style or material (galvanized metal, translucent panels, plywood with flashing and shingles, etc.)
- Two or three 8-ft (2.4-m) lengths of wide wooden dowels, 2x4s, or 2x2s, to use for the roosting bars (depending on the number of birds to be housed)
- Box each of 3½-in (90-mm) and 1⅝-in (40-mm) deck screws and washers
- Four metal brackets of chosen style
- Hinges: a set for one door, or two doors if also making a nest box
- Locks: one to three, depending on preference
- Wire mesh, approx. 6 x 16 in (15 x 40 cm)
- Wood glue (optional)
- Wooden or hardware handle (optional)

Tools

- Miter saw
- Circular saw
- Power drill
- Wire cutters
- Measuring tape
- Spirit level
- Square

Lumber dimensions

Two-by-four (four-by-two) and four-by-four—referred to throughout these projects as 2x4 (or 4x2) and 4x4—are nominal measurements used for standard cuts of lumber, indicating 2 x 4 in (50 x 100 mm), and 4 x 4 in (100 x 100 mm). Actual dimensions vary slightly, however, depending on location. In the US, for example, 2x4 measures 1½ x 3½ in (38 x 90 mm).

1 Measure, mark, and cut the coop walls from the siding, plywood, or OSB.

Tall (front) wall: 5 ft (1.5 m) high x 4 ft (1.2 m) wide
Short (back) wall: 3 ft (0.9 m) high x 4 ft (1.2 m) wide
Side walls: 5 ft (1.5 m) tall on the front edge x 4 ft (1.2 m) wide x 3 ft (0.9 m) tall on the back edge, creating a sloping roof

2 Mark and then cut out all three access doors.

Chicken-access door: cut out a doorway on the front wall panel, 13 in high x 8 in wide (33 x 20 cm). Save the cutout for a door reference if you'll be building your own sliding door (see page 284).

Clean-out door: cut out a doorway on one side wall, 32 in high x 32 in wide (81 x 81 cm). Save the cutout, which will become the clean-out door.

Egg-access door: cut out a doorway on the other side, 18 in high x 36 in wide (46 x 91 cm). Save the cutout to become the egg-access door.

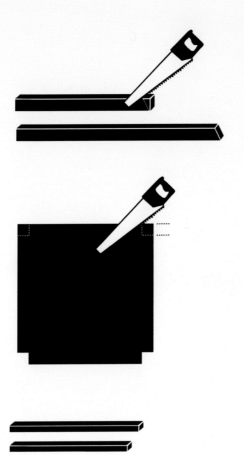

3 Mark and then cut the ventilation windows.

Cut a window, 3 in high x 6 in wide (7.5 x 15 cm), in the top-center of the front and the back wall. These should be located very close to the top edge.

Cover the back of these ventilation openings with hardware cloth (wire mesh), secured with the appropriate-sized screws and washers. Note: you can also choose to cover the windows with store-bought vent covers instead of hardware cloth.

4 Paint or stain the outside walls to your preference to protect the coop exterior. Since this is the exterior, any suitable paint or finish can be used without concerns regarding fumes.

5 Measure, mark, and cut the 4x4 support posts.

Front posts: 6 ft (1.8 m) high
Back posts: 4 ft (1.2 m) high

6 Miter the top of the four support posts so their angle matches the pitch of the side walls.

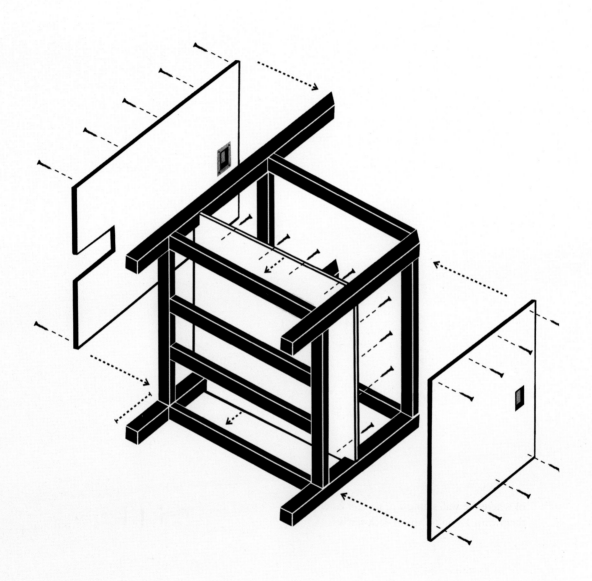

7 Cut and notch the coop floor.

Measure, mark, and cut a 4 x 4-ft (1.2 x 1.2-m) square from siding, plywood, or OSB.

Measure, mark, and cut 4 x 4-in (10-cm) notches in each corner so that the floor will fit around the four support posts.

8 Measure, mark, and cut support beams for the coop floor and walls.

Measure, mark, and cut eight 45-in (114-cm) support beams out of 2x4 to fit between the support posts: four for the top walls, four for the bottom walls.

Measure, mark, and cut two 40-in (102-cm) support beams out of 2x4 to fit between the front and back floor support beams.

9 Install the coop floor supports and the plywood floor by securing the floor through the siding and into the posts using 158-in (40-cm) screws.

With the coop lying on its side, attach four of the 45-in (114-cm) support beams between each support post at the floor level, 12 in (30.5 cm) from the bottom of the posts. Ensure these are flush with the bottom edge of where the wall siding will sit, and fasten them with 1⅝-in (40-mm) deck screws.

Attach the 40-in (102-cm) supports between the front and back floor support beams to provide additional support for the coop floor.

Carefully place the coop plywood floor between the posts so it sits flush on top of the floor supports. Fasten with 1⅝-in (40-mm) deck screws.

10 Add wall support beams between the coop's 4x4 post tops on all four sides. Ensure these are set flat against the coop's posts and flush with the tips of the mitered edges. Secure the support beams against the posts with toenailed screws going through the support beam into the 4x4 post. Toenailing involves drilling the screws through the support beams into the posts at an angle, used as common practice for secure structured frames.

11 Attach siding for three of the walls.

Align the front wall siding on top of the 2x4 support beams and align with the front two 4x4 support posts, then drive 1⅝-in (40-mm) deck screws through the siding and into the posts, evenly spaced along the length of the post. Repeat for the back wall siding.

Stand the coop upright and add the third wall siding. Don't add the fourth wall siding yet—you need to be able to access and finish the inside of the coop first.

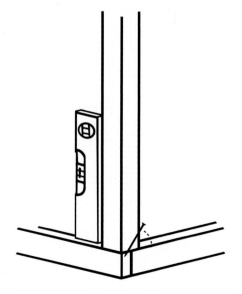

An example of toenailing

12 Build and install the nest boxes.

Cut two 12 x 12-in (30 x 30-cm) squares from siding or plywood. Attach the dividers vertically to a 2x4 rail, spaced evenly, using deck screws or wood glue. Make sure the dividers are flush with the bottom of the rail.

Attach the rail to the floor against the side that has the egg-access door. This creates three nest boxes.

If needed, cut and install a piece of plywood, 36 in long x 12 in wide (91.5 x 30.5 cm), to sit over the nest boxes to keep droppings off of the eggs.

13 Install the egg-access and clean-out doors.

Using the cutouts created earlier, align the doors to their respective openings, then attach them with the hinges.

Install your preferred locking mechanism on each door (if predators are present, you will want something more sturdy).

14 Attach the final wall siding, following the instructions in Step 11.

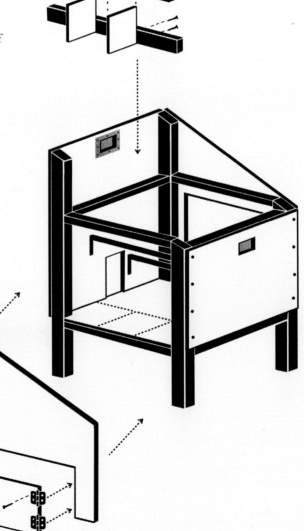

15 Install roosting bars.

Using sanded 2x4s, 2x2s, large dowels, or a combination of all of these, install two roosting bars with brackets. Place one roost a little higher than the other, ensuring that both roosts stay well below the ventilation windows. Double-check their placement with a level to ensure they are straight and will be comfortable for the chickens.

16 Cut and attach the roof.

Measure and cut the roof panels so they overlap with the front, sides, and rear of the coop: at least 4-in (10-cm) overlap on the sides and a 12-in (30.5-cm) overlap on the front and back.

Measure, cut, and miter 46-in (117-cm) 2x4 support beams that will sit flush with the front and back walls and support the roof panels. Fasten them in place with 3½-in (90-mm) deck screws.

Attach the roof panels to the coop.

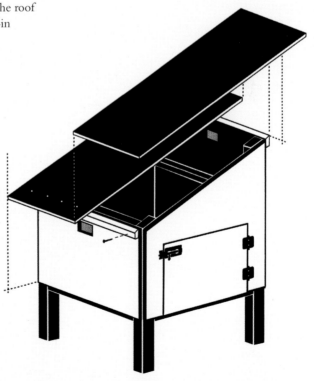

Build a sliding chicken-access door

This is an optional step, and only necessary if your coop is not located inside a secure run.

1. Using the cutout from the chicken-access door, measure and cut an additional rectangle that extends ½ in (1.3 cm) beyond each edge of the original cutout.

2. Cut two narrow plywood rectangles measuring 1½ x 15 in (3.8 x 38 cm) each, and two more rectangles measuring 2 x 15 in (5 x 38 cm) each.

3. Glue one smaller rectangle to a larger rectangle, making sure one side lines up flush, and the other side has a ½-in (1.3-cm) difference. Repeat with the remaining rectangles. These rails will allow the door to slide up and down while preventing it from falling.

4. Align the rails with the chicken-access opening, with the larger rectangle outward, and attach with deck screws.

5. Measure the space between the rails (approx. 14 in [35.5 cm]) and cut two identical rectangles that are 1½ in (3.8 cm) wide. Glue the two rectangles together, face to face. Attach them in the space between the rails using deck screws, with the bottom edge flush with the bottom of the opening, so the door does not slide out when closed.

6. Attach a small wooden or hardware handle to the bottom of the door.

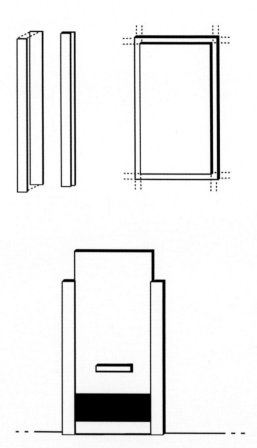

Customization ideas

The following ideas can help improve your coop, whether by giving it a fancier aesthetic, greater cleaning convenience, or easier access—for both humans and chickens.

- Add a small window to the clean-out access door for better air flow. Secure it with hardware cloth (wire mesh) and a hinged shutter.

- Replace the nest boxes and egg-gathering door with an external nest box to make gathering eggs easier. Free plans for these can be found online.

- Paint the walls on the inside of the coop to make them easier to clean and to prevent mites.

- Place linoleum on the coop floor for easy cleaning.

- Use fancy hardware for your coop hinges and handles.

- Paint the coop to match your house.

- Install signs, solar-powered lights, or other coop decor.

- Install an automatic-powered chicken-access door to help improve safety and temperature regulation without the need to open and close the door manually every day.

- Make the coop larger by extending the front and back walls to 6 ft (1.8 m) wide instead of 4 ft (1.2 m).

- Install a ramp from the chicken-access door down to the ground to make access easier for your chickens.

- Install trim on the coop's edges and doorways for a more finished look.

Make a chicken swing

Give your chickens a chance for some fun and exercise by creating a swinging perch. Keep in mind that chickens are rather heavy, clumsy birds, so the swing should be kept relatively low to the ground to encourage perching and prevent injury. Placed too high, a swing can cause a clumsy chicken to break a nail or get a puncture that could lead to bumblefoot (see page 214).

Materials
- One 8-ft (2.4-m) lumber board
- Narrow sisal rope: around ¼ in (6 mm) diameter
- Two carabiners (optional)

Tools
- Saw
- Sandpaper
- Drill

Opposite A busy chicken is a happy (and less destructive) one. Chickens love any opportunity to amuse themselves on swings, perches, gyms, and tunnels.

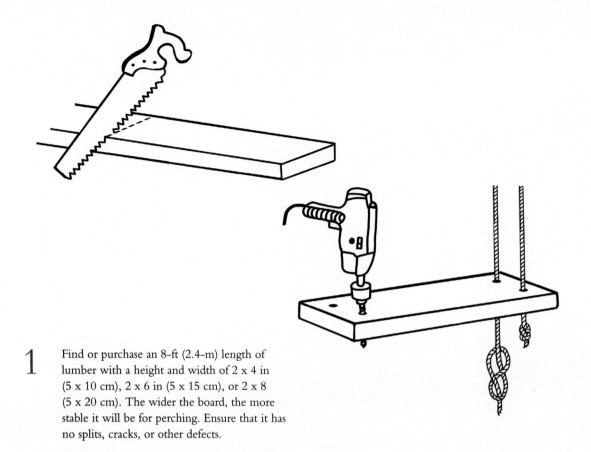

1   Find or purchase an 8-ft (2.4-m) length of lumber with a height and width of 2 x 4 in (5 x 10 cm), 2 x 6 in (5 x 15 cm), or 2 x 8 (5 x 20 cm). The wider the board, the more stable it will be for perching. Ensure that it has no splits, cracks, or other defects.

2   Cut the board down to your desired length—starting from a minimum of around 24 in (61 cm) long—and sand all sharp edges and corners to ensure the surface will be smooth and comfortable for your chickens.

3   Cut four equal lengths of rope, each long enough to suspend the swing about 12 in (30 cm) above the ground, and making sure you have an additional 12 in (30 cm) in reserve for tying and securing the ends. For your chickens' safety, don't hang the swing any higher than 3 ft (90 cm) above the ground.

4 Drill two holes through one side of the board, parallel to each other and about 2 in (5 cm) from the outer edge. The diameter of each hole will need to be big enough for your chosen rope to pass through. Repeat on the other side of the board, making sure the holes line up so that the swing will hang level. Take two pieces of rope and slip the ends through the drilled holes at one end, then tie a snug double knot beneath the board to hold them in place. Repeat the same thing at the other end of the board.

5 With all four lengths of rope now attached, tie the two ropes together on each side using a square knot about 12 in (30 cm) above the swing perch. This will keep the swing steady for your chickens.

6 Hang the swing in the run, making sure your birds have enough space to hop on and off. You can hang it directly from the sisal rope itself, or suspend it with the use of carabiners or similar attachments.

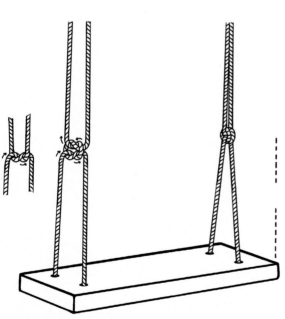

Make a chicken play gym

This fun A-frame play gym is a great boredom buster for backyard chickens. It fulfills their perching instincts and helps utilize vertical space in the run, making it especially useful for smaller spaces. It's also a very easy and affordable project. The suggested measurements can be adjusted to fit your flock's living space.

Materials
- Four 8-ft (2.4-m) lengths of 2x2
- Wood stain and sealant (optional)
- 3½-in (90-mm) deck screws

Tools
- Saw
- Sandpaper
- Drill

Opposite If you're low on budget or space, a chicken gym is an economical way to offer a climbing place for chickens and a fun DIY project for you.

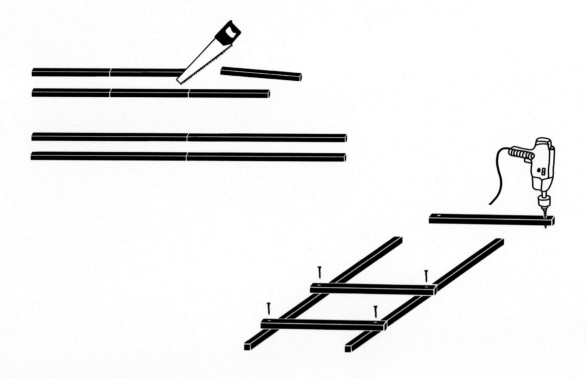

1   Find or purchase the lengths of 2x2, ensuring that they have no splits, cracks, or other defects. Using pressure-treated or finished lumber will increase longevity, but untreated and even natural wood is sufficient.

2   Measure, mark, and cut the lengths to size:

Four 4-ft (1.2-m) lengths for the A-frame structure

Six 2-ft (0.6-m) lengths for the rungs

3   Sand all of the rung edges down so that they'll be comfortable and safe for your chickens.

Optional: stain and seal the lumber to help it last longer. Because this is an exterior structure, with no concerns over fumes, all stains and finishes are suitable.

4   Assemble the first side of the A-frame.

Line up three of the rungs between two of
the longer support pieces, spaced evenly to
create a ladder-like structure. They should
overlap the supports a bit, making them
easier to attach.

Pre-drill holes to avoid splitting the wood,
then secure the rungs to the supports using
deck screws.

5   Repeat Step 4 on the second side to create a
second ladder-like structure.

6   Lean the two sides against each other so that
they interlock at the top. Adjust the angle to
your preference, then secure the two pieces
together with deck screws to form the
A-frame, again pre-drilling the holes to
prevent splitting.

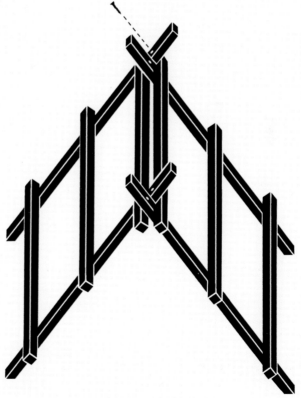

Make a chicken run

A backyard chicken run, or pen, may be an extravagantly built creation or simply a reused dog kennel, so long as it's reasonably sized and safe for a flock of chickens. This design creates a moderate 64 square foot (6 square meter) outdoor space, enough to comfortably fit six standard-sized chickens or nine bantams full-time, or more if free-range time is offered. It's also tall enough for a keeper to enter comfortably, and can easily enclose a small chicken coop. The corner braces of this run provide extra stability and add some aesthetic, too. The suggested measurements and materials here can be customized to fit your needs.

Dimensions
- Footprint: 8 x 8 ft (2.4 x 2.4 m)
- Height: 7 ft (2 m)

Features
- Fully enclosed wire pen
- Human-access door
- Predator skirt (optional but recommended)

Materials
- Four pressure-treated or stained/sealed 4x4 posts
- Twenty to twenty-five pressure-treated or stained/sealed 2x4s, each 8 ft (2.4 m) long (exact number may vary, depending on customizations such as the size of corner braces and the design of the door)
- ¼- to ½-in (6- to 13-mm) hardware cloth (wire mesh). Cattle panel or chicken wire may be used instead, but these materials are not as secure against predators.
- 3½ in (90-mm) deck screws and matching washers
- 1½ in (38-mm) deck screws and matching washers
- Hinges: for one door
- Locks: one or two, depending on preference
- Optional: roof panels, tarp, clear paneling , paint, coarse gravel, landscape staples

Tools
- Miter saw
- Circular saw
- Power drill
- Wire cutters
- Measuring tape
- Spirit level
- Square
- Shovel or pick

DISCLAIMER
Due to natural variance, lumber dimensions are not always precisely as labeled. Always double-check measurements with your own build before cutting. These measurements are to be used as a rough guide.

1   Measure, mark, and cut the four 4x4 posts to 7 ft (2 m) long. These will form the four corners of the chicken run.

2   Paint or stain the lumber to your preference to protect the chicken run from the elements. If you're using pressure-treated lumber throughout the run, this step is optional but still recommended.

3   Prepare the chicken run space.

    Using a hoe, shovel, or garden pick, level out the 8 x 8-ft (2.4 x 2.4-m) area where the chicken run will be constructed. Use a sturdy rake to spread the dirt evenly throughout the run, making sure the ground is as level as possible.

    If drainage is an issue, dig drainage ditches and line them with coarse gravel.

4   Determine whether the run will attach to the coop face (on the chicken-door side) or will entirely contain the coop within the run:

    If your run will contain the coop inside it, you may skip this step.

    If your run will attach to the coop itself, consideration must be taken to ensure a secure connection using the hardware cloth and 1½-in (38-mm) deck screws with washers to attach the cloth to the face of the coop. You will need to trim the cloth as necessary to secure the run.

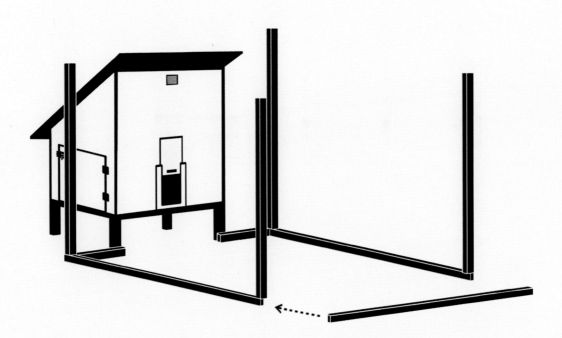

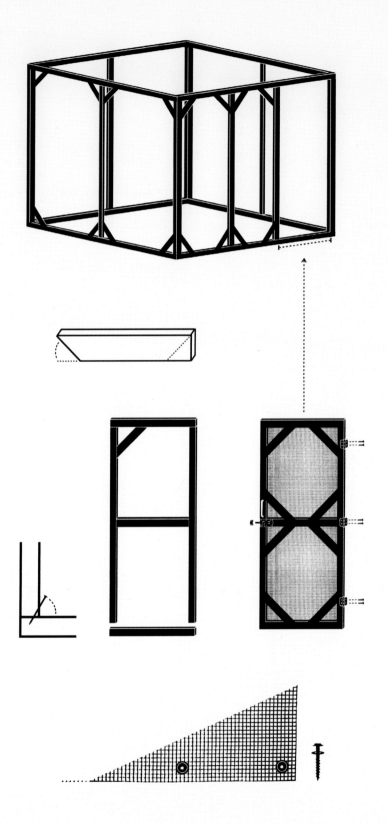

5

Place the initial 2x4s at the base of the run.

If you're attaching the run to the face of the coop, measure out two 2-ft (0.6-m) sections of 2x4 and attach them perpendicular to the forward legs of the coop using 3½-in (90-mm) deck screws on the left and right sides of the coop base. Attach the 4x4 posts with the same screws, using a spirit level to make sure they are perfectly vertical.

If you're enclosing the coop in the run space, place an 8-ft (2.4-m) length of 2x4 on the level ground and attach two 4x4 posts flush to the ends of the 2x4 board vertically, again checking their position with a spirit level.

Place additional 8-ft (2.4-m) 2x4s around the perimeter of the base, attaching each board flush with the 4x4 posts, and setting the posts vertically using a level to make sure they're straight. NOTE: Two sides of 2x4 will be 8 in (20 cm) shorter than the other sides, to compensate for the depth of the corners.

6

Set the 2x4s at the top rim of the run.

Using the same pattern and measurements as the run base, attach the 2x4s along the top flush edge of the run, connecting to each post flush with the edges. The height between the base and top edge should now be about 6 ft 4 in (1.9 m).

7

Measure out and mark 33 in (84 cm) from one edge of your preferred corner post.

Install the 2x4 support post at the mark to create the doorframe, using 3½-in (90-mm) deck screws.

8

Attach 2x4 studs as braces on three sides at each corner post.

Measure and cut six or seven studs measuring about 6 ft 4 in (1.9 m) from 2x4. Measure each individually, as the space can vary a bit.

Measure an equal spacing between the posts on each side, and attach using 3½-in (90-mm) deck screws diagonally through the corner of the 2x4.

If the run is attached to the coop face, this step will only apply to two sides.

9

To ensure the run structure remains strong and stable, measure, mark, and cut about four pieces of 2x4 at a 45-degree angle to act as additional corner braces for each 2x4 corner.

Right Paint or stain the lumber to your preference to protect the chicken run from the elements and to help it last longer. If you're using pressure-treated lumber throughout the run, this step is optional but still recommended.

10 Measure and cut two 68-in (173-cm) long 2x4s for the two sides of the door and two 30-in (76-cm) long 2x4s for the top and bottom of the door.

Measure and cut one 23-in (58-cm) long 2x4 for the center of the doorframe.

Assemble on a flat surface and connect the four doorframe sides together with screws.

Place the 23-in- (58-cm-) long 2x4 across the center of the doorframe, between the left and right sides. Secure with decking screws.

Measure and cut eight approximately 14-in (35.5-cm) long 2x4s at a 45-degree angle for the braces. Double-check measurements to ensure they fit snugly in the bottom doorframe.

With the door-frame still laying flat, install the wire fencing on the side facing upward. Secure with heavy duty staples or 1½-in (38-mm) deck screws with #10 washers.

Flip the door over, wire-side down, and install two or three hinges on one side using the hardware provided, or small screws. Install the door handle and external door latch on the other side.

Carefully fit the door to the run frame and secure with the door hinges. Attach the other side of the door latch to the run frame.

Optional: install extra hardware such as a door spring or internal door latch, as desired.

11 Cover the structure with wire fencing.

Open a roll of your preferred wire fencing and cut to the length you need. This should be either hardware cloth (¼ in to ½ in [6 mm to 12 mm] thick), cattle panel fencing, or chicken wire.

Secure the fencing with heavy-duty staples or 1½-in (38-mm) deck screws and washers (screws are far more secure than staples).

12 Install a predator skirt (optional).

Secure the perimeter of the run at ground level by installing a predator skirt using the wire fencing of your choice. This will prevent foxes and other predators from digging under the run. There are two ways of achieving this: digging in the wire fencing vertically beneath the ground, or laying it out flat on the ground and extending it outward.

Open a roll of your preferred wire fencing and cut it to at least 12 in (30.5 cm) wide, and to the length of each side of the run.

Option 1: dig a trench along each side of the run, 10 x 10 in (25 x 25 cm) deep, and install the wire fencing by securing it to the bottom 2x4s with deck screws and washers. Ensure that the fencing lays flat along the trench, curved outward. Backfill the wire-covered trench with rocks and soil.

Option 2: dig a wide, shallow area—around 1 to 3 in (2.5 to 7.5 cm) deep—extending at least 24 in (61 cm) out from the run. Lay the wire fencing flat against the ground and secure it to the bottom 2x4s with deck screws and washers. You can secure the wire fencing with landscape staples, or simply by placing heavy objects, such as cement blocks or large potted plants, on each corner.

Opposite A chicken run provides a safe space for chickens to roam and a wonderful opportunity to keep an eye on a beloved home flock. (As seen here, you can even decorate for the holidays!)

Make a chicken apron

Whether due to damage from an overzealous rooster or bullying from other hens, chickens can experience damage to the feathers and skin on their backs, particularly in the "saddle" area above the tail. This damage not only looks unsightly, it can cause discomfort and runs the risk of infection, particularly if the delicate skin gets broken. Fortunately, when this occurs, there is a relatively simple solution: fitting the chicken with a "chicken apron." Chicken aprons—sometimes called "hen saddles" or "saddle aprons"—can be purchased online, but they're also very easy to make at home with a scrap of fabric and a few basic sewing skills (either sewing by hand or using a machine).

Materials
- ¼ yard (meter) fabric in a pattern of your choosing (denim or cotton are recommended)
- ¼ yard (meter) soft fleece lining or batting (wadding)—lining may be skipped if a stronger fabric like denim is used
- Approx. 12 in (30 cm) narrow elastic, about ¼ in (6 mm) wide
- Fabric adhesive

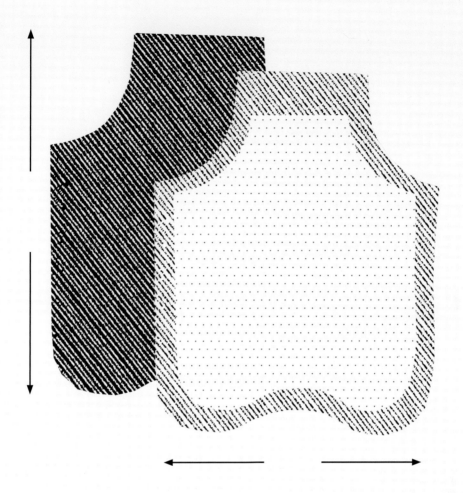

1   Cut two pieces of fabric in a "saddle" shape. Various patterns can be found online, or you can follow the pattern below.

Dimensions for a standard-sized hen should be about 9 in (23 cm) long and 8 in (20 cm) wide, with the top corners cut out for the chicken's wings, and the bottom edge cut out slightly to accommodate the tail. Exact dimensions may vary, though, depending on the chicken, and how extensive their feather damage is. It's best to "fit" everything first, to ensure you have enough fabric and elastic.

2   Cut the fleece lining or batting in the same shape, but smaller—there should be a 1- to 2-in (2.5- to 5-cm) margin of fabric around the fleece lining when laid flat.

3   Attach the fleece lining to one piece of the fabric, on the reverse side, using a fabric adhesive.

4   Assemble the pieces of fabric, with wrong sides facing out, and place the ends of the elastic under each curved corner at the top so that the elastic will sit just under and behind the wings. Pin everything in place.

5 Stitch the fabric together along the two sides and bottom of the pattern, leaving the top open.

6 Turn the apron right side out, and position the middle of the elastic so that it rests just under the top (neck) of the apron.

7 Fold the neck of the apron over the elastic, ensuring the two sides are symmetrical, then pin it in place.

8 Stitch the fold so that it attaches securely to the elastic.

9 You can add any simple embellishments you fancy to the apron, but avoid using anything that could come loose, such as ruffles, tulle, or bows, as these can become tangled and lead to injury.

10 Carefully fit the saddle to your chicken, taking extra care to ensure the elastic is pulled over the wings. The saddle should fit snugly, but not be too tight or uncomfortable. If the chicken pecks repeatedly at the saddle, shakes out its feathers, or holds its wings out after the first few minutes, consider checking the fit of the saddle and adjust it where necessary.

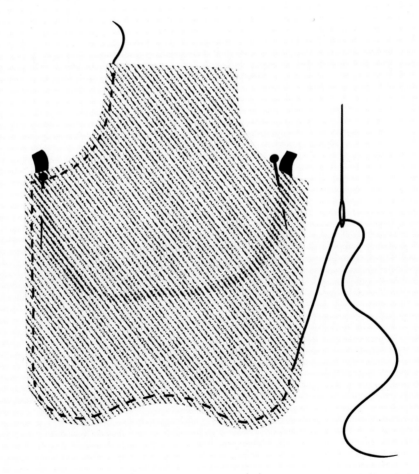

Above A well-fitting apron, or saddle, can speed up a chicken's recovery from feather loss or damage. **Following spread** Happy, healthy chickens are the result of choosing the right breed and supporting the best environment for you and your flock.

CHAPTER FOUR

THE ART
OF
CHICKENS

THE ART OF CHICKENS

Despite being the most populous group of birds on the planet, domesticated chickens (*Gallus gallus domesticus*)—thought to number around 23 billion on the planet at any one time—were relative latecomers to the evolutionary scene. Archaeological and genetic evidence appears to suggest that they originated simultaneously at multiple sites across the globe, including parts of China, South Asia, and Southeast Asia, most likely as a hybrid of the Red Junglefowl (*Gallus gallus*) and the Gray Junglefowl (*Gallus sonneratii*), and possibly the Green Junglefowl (*Gallus varius*) and the Sri Lankan Junglefowl *(Gallus lafayettii)*, all of which can still be found running wild today (see page 26).

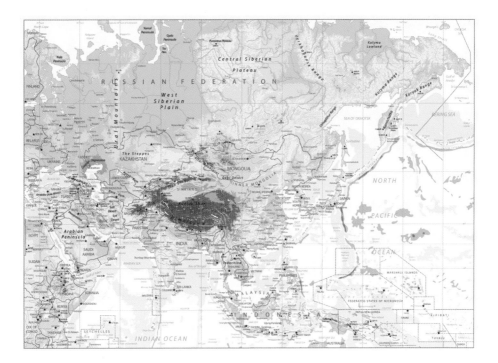

Wherever the domesticated chicken roamed, so chickens in art followed, serving as an illustration not just of their increased numbers but also new breeds, varied uses, ways of caring for them, and numerous symbolic associations relating to the mother hen, the protective rooster, and

Previous spread *Cockerels*, 1909 painting by Edgar Hunt. **Opposite** The Red Junglefowl (*Gallus gallus*) is understood to be the link to the modern-day chicken. **Above** Domesticated chickens appear to have originated at multiple sites, including across Asia.

the life-giving egg. Depictions of chickens in a diverse array of settings also provide an insight into the social, economic, and political landscape of humans through history and around the world. (Indeed, without chickens, for instance, the medium of photography may not have developed as it did, as the most popular nineteenth-century printing technique required coating paper with albumen—otherwise known as egg white; see page 359.) Such depictions have included chickens observed in their natural habitat (see Junglefowl, page 331), the barnyard (see page 336), at poultry shows (see page 369), and in modern gardens and homesteads (see page 391).

Chickens in art are also surprisingly widespread, from Benin Bronzes, the Chinese zodiac, and rooster- and egg-inspired folk art worldwide, through to illustrations for poultry-breeding manuals by the likes of Harrison William Weir, Joseph Williamson Ludlow, and Franklane Sewell (see page 232); paintings by master artists Hokusai (see page 326), Melchior d'Hondecoeter (see page 336), Pablo Picasso (see page 343), Marc Chagall (see page 349), and Gustav Klimt (see page 346); and the exquisite first Imperial Egg by Fabergé (see page 383). Chickens have also made their way into contemporary culture via works such as Koen Vanmechelen's Cosmopolitan Chicken Project (1999–ongoing; see page 362), Katharina Fritsch's Fourth Plinth installation *Hahn/Cock* (2013; see page 340), the lithographs and collages of Mark Hearld (see page 392), West African–influenced fabric designs by Vlisco (see opposite), and the animated movie *Chicken Run* (2000). Although never as prolifically portrayed as flowers, chickens even experienced a period of "Hen Fever" in the 1800s, which saw chicken prices soar to crazy heights, just as they had with tulips two centuries before.

Most of the artworks that exist appear to be of cockerels and roosters, owing to their generally more flamboyant appearance and associations with fighting, virility, loyalty, and perceived ability to ward off evil spirits by crowing at dawn. Hens, however, are now holding their own, as a renewed interest in chicken keeping (often of small, rooster-less flocks) has led to increased knowledge-sharing and image-making, with an interest in the pecking order when roosters are not around; breeds such as Brahmas, Cochins, Frizzles, and Silkies—the females of which are just as showy as the males, as witnessed in poultry-posing photography books by Moreno Monti and Matteo Tranchellini (see page 365) and Stephen Green-Armytage (see page 366); and a rainbow assortment of cream, green, pink, blue, and even chocolate-brown eggs.

If you're already passionate about chickens, their celebration through art will further compound that joy. If you're new to the coop, the wonderful pictorial narrative of these extraordinary birds and their ancient connection to human civilizations cannot fail to strike a chord.

Opposite (clockwise, from top left) The appearance of chickens in art has been surprisingly widespread, as seen here: from the West African–influenced fabric designs of Dutch textile manufacturer Vlisco (seen here in the iconic "Happy Family" pattern) to rural weathervanes across the globe to the very first Faberge egg given by Tsar Nicholas II to his wife, Empress Alexandra, in 1885.

ORNITHOLOGICAL ART
AND ILLUSTRATION

The fowl clade known collectively as Galloanserae is ubiquitous among the illustrated ornithological odes of natural history, including geographically diverse depictions of ground-feeding landfowl (galliforms such as pheasants, partridges, peacocks, and guineafowl), aquatic waterfowl (anseriforms such as ducks, geese, and swans), and poultry (including turkeys, quails, and chickens). While the humble domesticated chicken and its ancestors—notably the Red Junglefowl (*Gallus gallus*)—are certainly part of this oeuvre, there are eras when they really come to the fore, from symbolic Chinese and Japanese art to the full-color prints of mid-nineteenth-century Hen Fever that now proudly roost on many a country kitchen wall.

Opposite *Rooster, Hen, and Chicken with Spiderwort* by Katsushika Hokusai, *c.* 1830–33. **Above** *Rooster and Hen* by Ohara Koson, *c.* 1900–33.

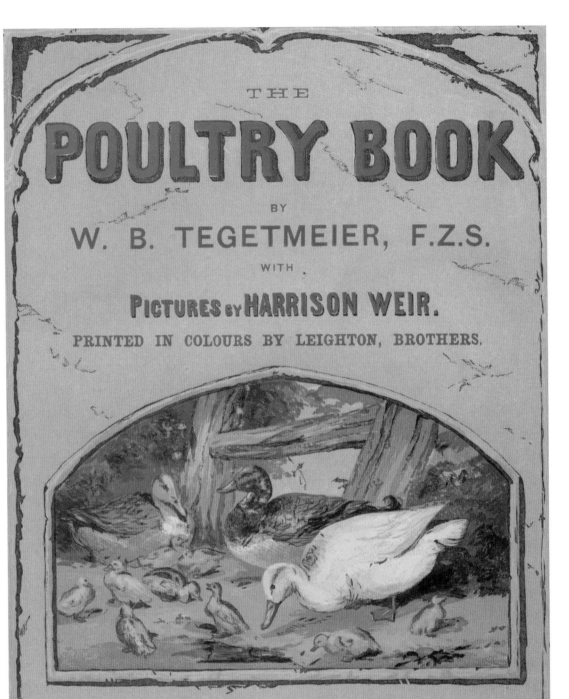

THE

POULTRY BOOK

BY

W. B. TEGETMEIER, F.Z.S.

WITH

PICTURES BY HARRISON WEIR.

PRINTED IN COLOURS BY LEIGHTON, BROTHERS.

LONDON:

GEORGE ROUTLEDGE AND SONS,

THE BROADWAY, LUDGATE.

NEW YORK: 416, BROOME STREET.

Hen Fever

(1845–55)

It's widely agreed that Queen Victoria was responsible for instigating the curious decade-long craze known as "Hen Fever" in 1842, when her established menagerie of exotic birds and beasts was supplemented with a brood of seven "Cochin China Fowl." Statuesque, with slender legs, long necks, rust-colored feathers, and a green-black tail, these showy birds completely eclipsed the drab farmyard chickens common at the time.

Gifted by a widely disliked globe-trotting sea captain by the name of Edward Belcher, whose logbook records picking up some poultry in Sumatra, the five hens and two roosters (retrospectively thought to be Shanghae or Malay chickens) were installed in the royal poultry house and subsequently bred with the Dorking—the ruler of the British roost since Roman times. The hope was that the offspring of these large, prolific egg-producing chickens might help alleviate the fallout of the Irish potato famine.

By 1849, descendants of those chickens had found passage to the east coast of America, where the popularity of poultry—and the fancying and exhibiting thereof—had quickly risen to great heights, culminating in the Boston Poultry Show of the same year. The obsession with owning and breeding the world's finest chickens, mirrored on both sides of the Atlantic, led to the publication of several now-classic illustrated manuals.

The Poultry Book by W.B. Tegetmeier (George Routledge & Sons, 1867) provided "The Standard of Excellence in Exhibition Birds" and was illustrated by Britain's preeminent animal artist of the time, Harrison William Weir (1824–1906). Although widely known as the "Father of the Cat Fancy," having organized the first ever cat show in England, Weir was also a keen fancier of poultry (inspired by

his parents and homestead life in the Kent and Sussex countryside), and he went on to write and illustrate *Our Poultry and All About Them* (Hutchinson and Company, 1903).

The Illustrated Book of Poultry by Lewis R. Wright (Cassell, Petter & Galpin, 1st edition, 1873; reprinted several times until 1911) was illustrated by ornithological artist and domestic-bird specialist Joseph Williamson Ludlow (1840–1916). First published in parts in 1872, it included a comprehensive series of fifty prints of prize British roosters, hens, and other fowl, portrayed in a farm or henhouse setting and with the aim of showing the temperament of the bird as well as its appearance. It also set a precedent for later editions of *British Poultry Standards*—now in its seventh, fully photographic edition (Wiley-Blackwell, 2018)—while the *Standard of Perfection* (1874), published by the American Poultry Association (founded in 1873), took the standard for poultry in the United States into the twentieth century. The latter was variously illustrated by leading poultry artists of the day, such as Franklane Sewell (1866–1945; also of *The Poultry Manual*, 1898), Arthur Schilling (1882–1958), Louis A. Stahmer (1873–1938), Irvin Burgess (1880–1920), Henry Lee (1861–95), and, more recently, Katherine Plumer.

Although hen hysteria self-combusted in 1855, due to rising prices and the pressures of high-maintenance henhouse regimes—as satirized in words and pictures in *The History of the Hen Fever: A Humorous Record*, by self-confessed chicken fancier George Pickering Burnham (James French and Company, 1855)—the many-feathered Heritage Chicken breeds of today, the "Standards" they must still live up to, and their painterly ancestral portraits live to tell the tale.

Opposite *The Poultry Book* by W. B. Tegetmeier, 1867 was an iconic celebration of animal artists of the time. **Following spread** A full-page illustrated article in the *Illustrated London News* from December 23, 1843, chronicling Queen Victoria's love for chickens and the burgeoning "Hen Fever" and, appearing over forty years later, "The Great Egg Question: Sketches at a Poultry Farm" from *The Illustrated London News* from April 2, 1887.

HER MAJESTY'S POULTRY HOUSE, HOME PARK, WINDSOR.

THE QUEEN'S POULTRY, WINDSOR.

A "good capon" and a "fatted cock" are, proverbially, the "graces" of the Christmas table; and, accordingly, the London markets have through the week teemed with the products of the country poultry-yards, and in many cases, such has been the demand, they have been overwhelmed with consignments from the French markets. The skins and hides of Leadenhall have fled before the "favourite article;" and even the umbrageous stalls of Covent-garden have, in one instance, assumed the *alias* of a poultry shop. The subject, since we are taught, "with reason," "to eat" as well as "to admire," is, therefore, a popular one. The ready-to-be-roasted chicken forms an apt prelude, and by forethought a very savoury one, to the profitable mysteries of the POULTRY-YARD; and, as we were some time since favoured by her Majesty with an inspection of her

elegant new poultry-house and its unique collection of fowls, and were at the same time honoured with permission to make such sketches as might be necessary for their adequate illustration, we are tempted to anticipate the vernal season, when we had intended to publish them, and now, at a season when the "thrice crowing cock" himself is wearied in summoning his family to daylight, we have re-solved to show to our readers a fair and seasonable picture of our gracious and nature-loving Queen, and of

The habitations and the little jo, s

of her winged favourites in the royal poultry yards of Windsor.

In a secluded nook on the boundaries of the Home Park, sheltered from the prevailing winds by stately clumps of elm trees, stands the HOME FARM—or the farm attached to Windsor Castle—the private farm of her Majesty. In this establishment, which was founded by

George III., is situated the royal owl-house and poultry-yards, which we have engraved at the head of this article, but of which, notwith-standing their great interest, the public know nothing, save the mere fact of their existence. Here her Majesty, retiring from the fatigues of state, finds a grateful relief in the simple pursuits of a country life; and here, too, it may be, like Louis XVI. in the Jardin Anglaise of the Petite Trianon, she seeks the renovation of those higher powers which find their best, if not their only home, in ∫nature, or its God. In cultivating the homely recreations of a farm, her Majesty has exhibited great industry and much good taste. The buildings and the farm routine which sufficed for the clumsy management of 1793, have been discovered by her Majesty to be totally unsuited to the more enlightened system of 1843, and hence, under the direction of her Majesty and Prince Albert, assisted by Major-General

HER MAJESTY'S COCHIN CHINA FOWLS.

1. Field with poultry coops.
2. Specimens of good breeds.
3. Incubators, for artificial hatching.
4. Artificial foster-mother, for newborn chickens.
5. Sleeping-house for older chicken.
6. Fattening chicken with the "cramming machine."
7. Higgler and boy collecting poultry.
8. Coops in barn, with chicken fattening.
9. Plucking chicken (killed by the man).
10. Chicken being pressed for packing.
11. Packing-cases for chicken.

THE GREAT EGG QUESTION: SKETCHES AT A POULTRY FARM.

Chinese and Japanese art
(10th–20th century)

While nineteenth-century Hen Fever was inspiring some of the most charming and realistic portrayals of domesticated chickens in the West, Chinese and Japanese artists were continuing their centuries-old tradition of painting farmyard hens and roosters (the latter being the tenth of the twelve animals of the Chinese zodiac)—as symbols of motherhood, domestic harmony, military prowess, and good fortune.

Although early theories, including those asserted by Charles Darwin, suggested that chickens were first domesticated in the Indus Valley some 4,000 to 6,000 years ago, new evidence suggests that they might have been reared in northern China as far back as 10,000 years ago, when its climate was warm enough to host the Red Junglefowl (*Gallus gallus*). Chickens were most certainly part of the cultural fabric of Chinese life during the Song (960–1279), Yuan (1271–1368), and Ming (1368–1644) dynasties, as evidenced by artistic works by Wang Ning (eleventh century) and Emperor Xuangzong (1399–1435).

Writings show that chicken and eggs were both on the menu by this time, thanks to artificial incubation techniques (hot-housing eggs, freeing up hens to lay more) and evolving farming practices. There was a pecking order in place at the table, however, running in parallel with the ancient pastime of cockfighting, with prize birds specially bred for royalty and the wealthy as a symbol of power and prestige. The rooster as cock and king was still very much a theme into the Qing dynasty (1644–1912) and beyond, found proudly strutting across the prints and paintings of Xu Beihong (1895–1953), Ju Chao (1811–65), and renowned bird-and-flower artist Yang Shanshen (1913–2004).

Chickens were also a favored subject matter of Japanese painters and printmakers, including the artist and Zen Buddhist Ito Jakuchu (1716–1800), who learned to observe animals as a child in the marketplace; Ohara Koson (1877–1945), who created numerous elegantly drafted *kacho-e* (birds-and-flowers) woodblock prints in the *shin hanga* ("new prints") style; and the *Great Wave* painter Hokusai (1760–1849). As an extra adage, it's said that Hokusai was once called before the Shogun's court to demonstrate his artistic prowess. He painted a long blue mark on a sheet of paper, then chased a chicken across the page—its feet having first been dipped in red paint—to create the effect of autumnal leaves floating on the Tatsuta River. The chicken as artist and muse.

Above *Chicken and Rooster* by Ohara Koson (precise date unknown). **Opposite** *Rooster Hen with Hydrangeas* by Ito Jakuchu (precise date unknown). **Following spread** This stunning screen chronicling the life of a chicken was also illustrated by Ito Jakuchu. Both artists were prominent Japanese painters and printmakers who favored chickens in their subject matter.

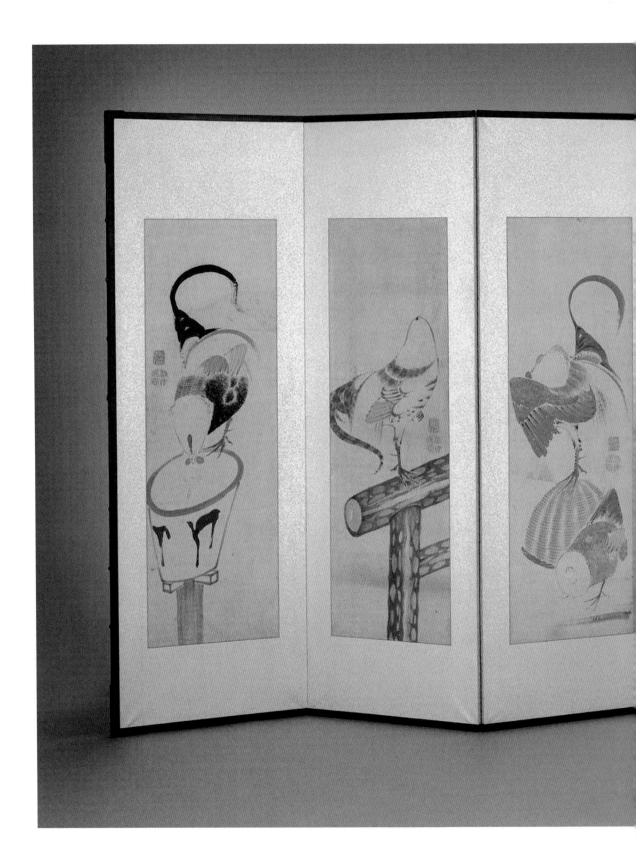

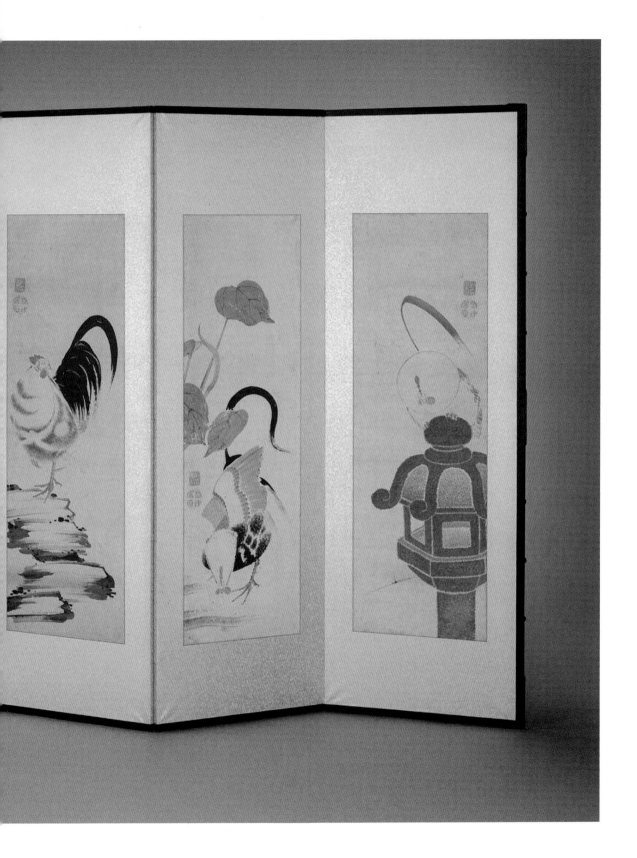

Sonnerats Hahn.
Gallus Sonnerati.
Coq Sonnerat.

¼

Sonnerats Henne.
Pule Sonnerat.

Der Hahn von Bankiva
Gallus Bankiva
Coq Bankiva

¼

Der Kupferfarbne Hahn.
Gallus aeneus
Coq bronzé.

Junglefowl

While the riddle about whether the chicken or the egg came first is ongoing, it's widely agreed that the Red Junglefowl (*Gallus gallus*, native to South and Southeast Asia) is the main progenitor of the domesticated chicken, with some suspected "fowl play" (natural hybridization) from the Gray Junglefowl (*Gallus sonneratii*, endemic to India), the Green Junglefowl (*Gallus varius*, endemic to Java and nearby islands), and the Sri Lankan Junglefowl (*Gallus lafayettii*, endemic to that country). Although all four species still run wild today, it can be quite hard to catch a glimpse of these shy birds, despite the males' showy plumage and bright red comb and wattles, in contrast to the dowdy but easily camouflaged coloring of the females. It's fortunate, then, that these charismatic creatures were captured by some of the best bird observers in the nineteenth century, including Swiss illustrator and lithographer Karl Joseph Brodtmann (1787–1862), ornithological artist John Gould (1804–81), sculptor, natural-history artist, and renowned dinosaur model-maker Benjamin Waterhouse Hawkins (1807–94), and prolific illustrator and falconry expert George Edward Lodge (1860–1954)—the latter producing a whole series of junglefowl portraits for the New York Zoological Society's *A Monograph of the Pheasants* by William Beebe (1918–22).

Opposite *Red Junglefowl* (1831–33) by Swiss illustrator and lithographer Karl Joseph Brodtmann shows how enthusiastically these birds were captured by eighteenth-century artists. **Above** *Javan Hen (Gallus furcatus)* by Benjamin Waterhouse Hawkins (precise date unknown).

Francis Barlow

(c. 1626–1704)

Fowl crop up numerous times in the tales of the Greek (or some think Ethiopian) fabulist and storyteller known as Aesop (*c.* 620–564 BCE). Although it's not certain whether Aesop was himself fact or fiction—no manuscripts by him survive and details of his life are scattered—these clever, animal-orientated, allegorical narratives were somehow handed down through the centuries to become one of the first illustrated books to be printed, first in Germany in 1461, then in English by William Caxton (*c.* 1422–*c.* 1491) in 1484. Many versions followed, including a 1665 (and a revised 1687) English, French, and Latin edition by the prolific painter and illustrator Francis Barlow. Widely regarded as Britain's first wildlife painter, the "Father of British Sporting Painting," and a pioneer of comics, Barlow produced the original drawings for the book, as well as the etching of the plates. Renowned for his attention to detail in the portrayal of animals, using graphite, ink, and watercolor, his influence can also be found in numerous "after Francis Barlow" works, such as those by the Bohemian graphic artist Wenceslaus Hollar (1607–77), who had illustrated a rival *Fables of Aesop* by the Scottish translator and cartographer John Ogilby in 1665, and went on to illustrate Ogilby's follow-up, *Aesopics*, in 1668.

So esteemed was Francis Barlow for his wildlife paintings and etches, he influenced dozens of prominent artists in his wake. **Below** *A Cock, Hen, and Six Chicks in a Farmyard* (after Francis Barlow) from "Diversae Avium Species" by Wenceslaus Hollar (1654-58). **Far left** *The Cock and the Precious Stone* from "Aesop's Fables" by James Kirk (1760).

PAINTING AND SCULPTURE

From bucolic portrayals of barnyard scenes (Melchior d'Hondecoeter and Walter Frederick Osborne), painterly reveries of roosters (Marc Chagall and Gustav Klimt), and pop art nods (Ed Ruscha and Andy Warhol) to culturally displaced Benin Bronzes, zodiac references (Ai Weiwei), and giant blue cockerels (Katharina Fritsch), the domesticated chicken populates the works of some of history's most prominent and influential artists and sculptors. While the male of the species receives the most overtures—due to its decorative plumage, intriguing comb, wattle, and spurs, plus centuries of cultural associations with courage, vitality, and fertility—hens, chicks, and eggs do appear to address the balance of life.

The incredible breadth of chickens portrayed in sculpture and painting is astounding. **Above** Joana Vasconcelos' supersize *Pop Galo* ("Pop Rooster") of 2016 and the painting of Joseph Crawhall, demonstrated in this 1903 painting entitled *Spangled Cock* (**Opposite**).

Art of the poultry yard

(17th–20th century)

Chickens crop up in paintings throughout art history. Some eras and movements, however, are more amply flocked with domesticated fowl, with notable animal artists such as the seventeenth-century "Father of British Sport Painting" Francis Barlow (see page 332) and his contemporary, the Dutch painter Melchior d'Hondecoeter (1636–95), ruling the roost.

Although the "Golden Age" of Dutch painting is more commonly associated with history, portraiture, flowers, and still lifes, other categories such as landscape art and genre painting cast a net over a menagerie of fair fowl, from the roosters and hens of the barnyard to partridges, pigeons, and more glamorous relatives such as peacocks, posing in the grounds of palatial country manors. Melchior d'Hondecoeter covered all these birds and more, following in the footsteps of his grandfather Gillis (1575–80), father Gijsbert Hondecoeter (1604–53), and uncle Jan Baptist Weenix (1621–*c.* 1663), juxtaposing realistic depictions of exotic breeds and exuberant wildfowl to conjure up an idealized view of the farmyard. These were paintings designed to adorn the walls of wealthy patrons, eager to show how cultured they were.

Such works inspired fellow animaliers such as the Belgian Eugène Joseph Verboeckhoven (1798–1881), who not only painted and printed his own pastoral scenes but was also called upon to enrich the works of other artists with his deftly observed birds and animals—among them sheep, donkeys, and chickens—for which he would produce numerous preparatory works.

Animals and rustic scenes also occupied the less grandiose canvases of the English artist Edgar Hunt (1876–1953), who began sketching farm life as a child, encouraged by his uncle, fellow animal painter Walter Hunt (1861–1941). In parallel, the Irish Impressionist and Post-Impressionist painter Walter Frederick Osborne (1859–1903), who specialized in landscapes and portraits, was also laying down farmyard scenes, albeit with the focus on the day-to-day life of ordinary people—and in particular, Ireland's working-class poor.

The Belgian Eugène Rémy Maes (1849–1931) see page 334 and English-born Joseph Crawhall III (1861–1913)—a member of the Scottish art collective known as the Glasgow Boys, who championed rural, prosaic scenes—also devoted a significant part of their oeuvre to the painting of barnyard fowl. Maes created picturesque narratives complete with barns, fences, sheep, and dogs, while Crawhall, under the obvious influence of Chinese and Japanese art (see page 326), tended to prefer producing one vibrantly painted bird at a time.

By no means are these the only artworks from the seventeenth to the twentieth century devoted to the poultry yard, but between them they do give a feel for what living with chickens was like at the time: free-range, largely humble ground-feeders, living in apparent harmony with other animals in the yard, still sharing a connection with other wild and domesticated fowl, with the odd exotic breed drawing attention to an ever-more globally connected world.

Right *Feeding the Chickens,* by Walter Osborne (1885), portrays the day-to-day life of the working class in Ireland in an Impressionist style.

Okukor

(16th–17th century)

In 2021, a bronze cockerel known as *Okukor* flew free of its more than 100-year residence at Jesus College in Cambridge, England, and back to Nigeria, location of the former Kingdom of Benin. The cock, one of thousands of Benin Bronzes that once decorated the kingdom's royal palace, had been looted by the British during the punitive and bloody Benin Expedition of 1897 and was finally coming home to roost.

Created by artists of the Edo people, the Benin Bronzes included elaborately decorated plaques, commemorative heads, and animal and human figures; they not only represent some of Nigeria's most important historical and cultural objects—many of them are also imbued with symbolism. Bronze cocks such as *Okukor*, for example, are thought to have been ceremonial items, displayed on the ancestral altar of an *iyoba* (Queen Mother), the senior wife of the *oba* or traditional ruler. The male of the species was chosen to signify the power and privilege of such women, who were given the honorific title "Eson, Ogoro Madagba," translating as "the cock that crows at the head of the harem:" the cock that sings the loudest.

Made using the ancient lost-wax process—whereby a metal sculpture is cast using a mold formed around an original wax sculpture—the Benin Bronzes (most actually made of brass or copper alloy) are renowned for their masterful production and highly sophisticated detail. *Okukor*, one of two dozen known bronze cocks (*eson*), is no exception, with its finely detailed feathers, feet, comb, and wattle. Such artifacts are also known to have inspired various art movements in the West, including Cubism, Fauvism, Expressionism, and Surrealism, as well as Pablo Picasso's so-called African Period, which lasted from 1907 to 1909.

In present-day Benin City, bronze cocks crow on, thanks to a decree by Oba Eweka II (1914–33) allowing artists of the brass casters guild to make sculptures for the wider market as well as for the royal court. It's hoped that all the Benin Bronzes will return home to join them soon, to be jointly housed in a dedicated museum.

Opposite This bronze cockerel, likely from the eighteenth century, is one of the "Benin Bronzes" created by the Edo people. It was probably used to decorate a shrine at the court of the Queen Mother in Benin.

Katharina Fritsch

(1956–present)

When German artist Katharina Fritsch first considered putting forward an artwork for the Trafalgar Square Fourth Plinth Commission in London, her idea was quick to develop. Inspired by a stuffed rooster in her studio and the role of a male posturing in the square, the concept for *Hahn/Cock* was born.

Taking the form of a giant glass-fiber polyester resin cockerel, 15½ feet (4.7 meters) tall and painted in an arresting shade of ultramarine blue, the sculpture proudly took up its place on the Fourth Plinth in 2013—the ninth artwork displayed in this globally renowned space since the initiative launched in 1999—where an equestrian statue of King William IV was once destined to stand. (It remained bare due to a lack of funds at the time.) The rooster's fellow statues included King George IV in Roman attire upon a horse (erected on the northeastern plinth in 1843), tributes to Charles James Napier (southwest; 1855) and Henry Havelock (southeast; 1861), and of course, Nelson's Column (complete with four bronze lions at its base), built to commemorate Admiral Horatio Nelson, who died at the Battle of Trafalgar in 1805.

Although Fritsch's choice of subject matter was overtly male, she saw it as a feminist piece, a commentary by a female artist on public shows of male power, pomp, and success. When the statue was erected—to much headlining innuendo, due to its nature and name (*hahn* and "cock" are slang for the same thing)—people were left wondering whether there was also a connection to France's national animal, the Gallic rooster, and by association, to France's former military and political leader (and Nelson's legendary adversary) Napoleon Bonaparte (1769–1821). The artist said not, but the humor was not lost.

In terms of visual impact alone, the giant blue rooster provided a startling and fun pop of bright blue for visitors from around the world, its flamboyant ruffled tail, erect crest, puffed-out breast, and jutting spurs bringing an animalistic vitality to this urban space (pigeons not included). Having completed its stint on the Fourth Plinth in 2015, followed by a temporary residence at the private art museum Glenstone, in Maryland, the big blue bird now lives plinth-less but illuminated by night on the roof terrace of Washington's National Gallery of Art, where it came to serve as a symbol of hope, renewal, and collective American endurance during the Covid-19 pandemic.

Opposite The giant blue glass-fiber polyster resin *Hahn/Cock* by Katharina Fritsch stands at over 15 feet (5 meters) tall, and in 2015, it held pride of place on the Fourth Plinth in Trafalgar Square, in London, England.

Pablo Picasso

(1881–1973)

It's no surprise that the Spanish artist Pablo Picasso included paintings and drawings of cockerels in his work. An animal imbued with the symbolism of war, vitality, and sexual prowess, such imagery aligned perfectly with his artistic drive and personal experiences throughout the Spanish Civil War (1936–39) and World War II (1939–45), the era during which many of his rooster works were created.

What set Picasso apart from the many others throughout art history who had depicted the birds in their work was the sheer range of approaches he used—from graphic single-line drawings (1918) and the black-and-white etchings he produced for the *Texts of Buffon* (1936, published 1942) to his brightly colored oil paintings of roosters of 1938 and 1943.

Combining stylistic elements of Expressionism, Cubism, and Surrealism, and illustrating the facial contortions and dynamic movements of the bird as well as its physical appearance, one theory surrounding this "rooster series" is that these spurred birds represented America as it entered the war—signifying, perhaps, the dawn of a hopeful new era.

Picasso's friend, the French poet, designer, and artist Jean Cocteau (1889–1963), also ventured into the barnyard to produce his own *Rooster* in 1956, albeit quieter in color and less animated in stance. A different artist but also a very different time. A French documentary of Picasso released that same year—*The Mystery of Picasso*, directed by Henri-Georges Clouzot—suggests a similarly playful, post-war approach to his work, and indeed chickens, as a simple drawing created on a transparent canvas with a marker pen morphs from a bunch of flowers to a fish to a chicken to a face, each act obliterating the previous one. A re-creation of his *Visage: Head of a Faun* (1955), it is a five-minute masterpiece of art in the making, stroke by stroke.

Opposite *Le Coq* by Pablo Picasso (1938) portrays a staple subject matter of the Spanish artist's work from the time, that of the symbolic rooster.

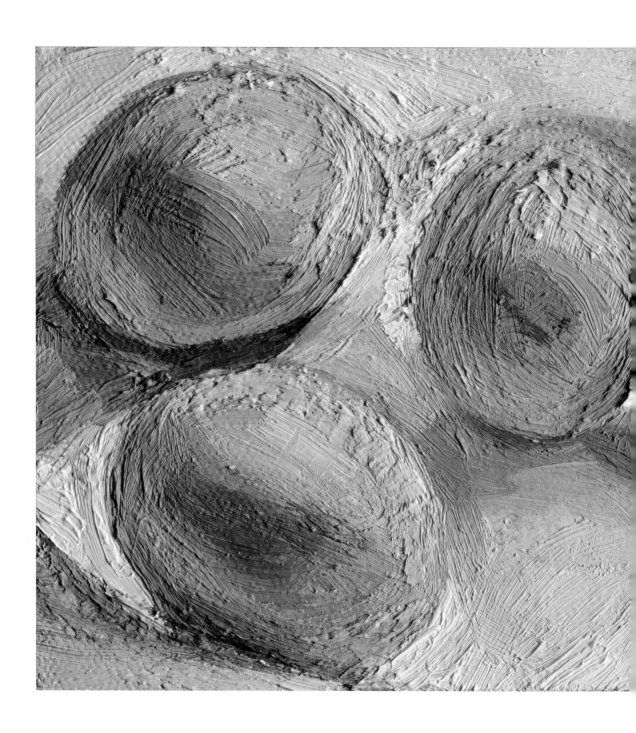

Lucian Freud

(1922–2011)

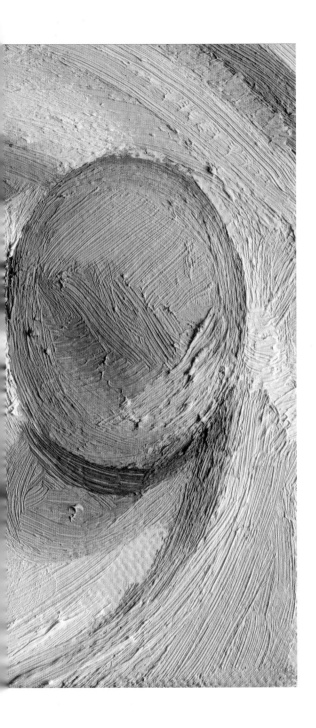

The British painter Lucian Freud is renowned for his figurative art and portraits, the sitter often naked, their flesh captured from life using an expressive paint-layering technique known as impasto. A postcard-size oil painting depicting four chicken eggs on a plate was given the same treatment, transforming a simple subject into something much more life-affirming and powerful. *Four Eggs on a Plate* (2002) was given to Freud's close friend the late Deborah Cavendish, neé Mitford, Dowager Duchess of Devonshire. The Duchess was a keen poultry enthusiast and would bring Freud freshly laid eggs every time she visited London from Chatsworth, the family seat in Derbyshire. According to Sotheby's, who sold the work upon her death, she kept the painting in a blue box with a note reading: "Box & rag he uses in his studio containing the painting of 4 eggs given me by Lucian Freud autumn 2002, DD." The Duchess herself was also painted by the artist, but the most famous portrait of her is a photograph by Bruce Weber, dressed in a ballgown and feeding her beloved chickens. Her love of these birds was also reflected in the titles of her memoirs: *Counting My Chickens . . . And Other Home Thoughts* (Long Barn Books, 2001), *Home to Roost: And Other Peckings* (John Murray, 2009), and *All in One Basket: Nest Eggs* (John Murray, 2012).

Opposite Lucien Freud's 2002 painting *Four Eggs on a Plate* was given to the artist's close friend and chicken enthusiast, the Duchess of Devonshire. The Duchess also appears on page 391, as a friend and fellow chicken appreciator of the contemporary artist Arthur Parkinson.

Gustav Klimt

(1862–1918)

Most famed for his "Golden Phase" (around 1898–1910)—characterized by paintings that used gold leaf, such as *The Kiss* (1907–08), as well as portraits of Adele Bloch-Bauer and other high-society friends and associates—the Austrian artist Gustav Klimt also produced a number of landscapes, including *Garden Path with Chickens* (1916). Painted while at his usual summer residence on the shores of Austria's Lake Attersee, this pastoral scene, with its proliferation of jewel-colored flowers, feels almost like a self-made escape from the turmoil of World War I (1914–17) that was raging on around him. Acquired by August Lederer, a Jewish art collector and supporter of the artists of the Vienna Secession, the painting enjoyed a brief period of peace before being confiscated by the Nazis along with the rest of the Lederer Collection during World War II and transported to the Schloss Immendorf in Lower Austria for safe storage. On May 8, 1945, the last day of the war, the castle caught fire (presumably started by retreating German troops), and the painting was destroyed. Any sightings of the artwork since this time have therefore been reproductions based on photographic records that were thankfully taken before it was seized.

Opposite Gustav Klimt's landscape painting *Garden Path with Chickens* (1916) is based on an idyllic scene at the artist's summer residence. Tragically, the original painting was lost in a fire on the last day of World War II.

Marc Chagall

(1887–1985)

The colorful, childlike dreamscapes of French modernist Jewish artist Marc Chagall are known for a range of symbolic motifs, including cows, fiddlers, and flying people. Roosters are another frequent part of the mirage, often as part of a scene of amorous love, the crested bird representing notions of fertility, nurture, and vitality. The rooster also stands proud as part of Chagall's most enduring theme, the spirit and folklore of his beloved childhood home Vitebsk, in Belarus (then part of the Russian Empire), where he was born Moishe Shagal to a Hasidic Jewish family and lived until leaving first for St. Petersburg in 1907, and then Paris in 1910. Chagall survived several periods of profound personal and political turmoil. While on a visit back home to see his new fiancée, the writer Bella Rosenfeld, World War I broke out and the Russian border was closed. The couple then remained in Vitebsk until 1923, living through the Bolshevik Revolution of 1917 and the regime change that followed, during which period Chagall founded an art school. Later, living in France during World War II, the Chagalls fled the Nazis by escaping temporarily to the United States, but in 1944, just as Paris was liberated and they were contemplating their return, Bella died from a bacterial infection due to a lack of medicine. Despite all this, the spirit of Chagall's work remains somehow poetic and surreally harmonious—with perhaps the rooster, that avian symbol of resilience, forgiveness, and selfhood, keeping him going through it all.

Opposite Modernist painter Marc Chagall frequently included roosters in his works. *The Bridal Pair with the Eiffel Tower* (1939), replicated here on an official French stamp in 1963, uses the symbol of the rooster to show love, vitality, and fertility.

Joana Vasconcelos

(1971–present)

In 2020, the internationally renowned Yorkshire Sculpture Park (YSP) near Wakefield in England staged *Beyond*, a major body of works by leading conceptual Portuguese artist Joana Vasconcelos. Among these was the supersize chicken *Pop Galo* ("Pop Rooster"; 2016). Inspired by the Galo de Barcelos, or Rooster of Barcelos—from a medieval Portuguese legend about a dead rooster that saves an innocent man from punishment for a crime he didn't commit, and now one of the country's most popular icons and pottery motifs—the tiled sculpture stood 29½ feet high (9 meters), covered in thousands of glazed tiles. As dusk fell, some 15,000 LED lights incorporated into the tiles illuminated the piece as a composition by musician Jonas Runa played from within, a playful pop-art statement about the handcrafted versus the industrial. This wasn't the first place *Pop Galo* had made its home, having started its tour at the annual Web Summit technology conference in Lisbon in November 2016. From there it had flown to Beijing to celebrate the Year of the Rooster (April 2017), Bilbao in Spain as part of Vasconcelos's solo show *I'm Your Mirror* (June–November 2018), its legendary hometown of Barcelos (December 2018–September 2019), and finally the YSP (for an extended visit until January 2022). Wherever it flies next, the symbolism with which it is imbued will hopefully bring luck and happiness to all who see it.

Opposite The supersize sculpture of *Pop Galo* ("Pop Rooster"), in an English sculpture park, stands at nearly 30 feet high (9 meters), covered in thousands of glazed tiles and 15,000 LED lights.

Ai Weiwei
(1957–present)

Like Joana Vasconcelos (see page 350), the Chinese artist Ai Weiwei has also taken a rooster on a multiyear, worldwide tour, as part of his 2010 bronze series *Circle of Animals/Zodiac Heads*. This is an upscaled reinterpretation of a set of bronze time-telling fountainheads representing each of the twelve animals of the Chinese zodiac at Beijing's Summer Palace retreat—designed and produced by Jesuit missionaries Giuseppe Castiglione and Michel Benoist for the Qianlong Emperor (1711–99) but subsequently looted by the British and French in 1860 during the Second Opium War. Weiwei's sculpture seeks to convey ideas around cultural belonging and the impact of the expulsion, migration, and deliberate displacement of both people and objects. However, it also has a playful element, and its depiction of the Chinese zodiac is something we can all find a connection to, whether we follow it or not. The 800-pound (363-kilogram), 10-foot (3-meter) rooster, along with its compadres the rat, ox, tiger, rabbit, dragon, snake, horse, goat, monkey, dog, and pig, has traveled extensively since it was first revealed in 2010, in parallel with a smaller, gold version of the work. And of the original bronze animal heads, seven have now found their way back to China, including the rat and the rabbit, which turned up in the collection of late French fashion designer Yves Saint Laurent. Five, however, are still at large, including the elusive rooster.

Opposite Part of a touring series of bronze heads depicting the Chinese zodiac, Ai Weiwei created this rooster head that stands at 10 feet (3 meters) tall. In the zodiac, the rooster represents fidelity and punctuality, as well as resourcefulness, and is a symbol of luck.

Ed Ruscha

(1937–present)

The American artist Ed Ruscha has a habit of making the ordinary extraordinary, using a range of media, including collage, painting, typography, and photography, to elevate the everyday to art-world heights. It's no surprise, then, that the domesticated chicken, so synonymous with both backyard life and the get up and go of the American way, thanks to Kellogg's and its iconic Cornelius "Corny" Rooster mascot, features in countless of his works. First appearing in a gelatin silver print entitled *Ross the Rooster* in 1960—a precursor to the documentary-style photographic works with which Ruscha filled "on-the-road" artist books during the 1960s and 1970s—the farmyard bird was then reduced to a series of typographic words and egg-inspired works on paper before cropping up in perhaps their most recognized incarnation: the stenciled silhouettes of *Rooster* (1987), *Zip Rooster* (1994), *Father of a Chicken* (1997), *Father of Some Chicken* (1997), *Chick Unit* (2003), and *Little Chick* (2007). Although the subject matter is wholly simple, the combined effect is powerful—not unlike the work of fellow pop artist Andy Warhol, who found fame with his iconic *Campbell's Soup Cans* series (1962), which included a homage to the popular flavor "Chicken Noodle," a short hop from an earlier branded work entitled *Rooster with Coca Cola Bottle* (1960).

Right Multimedia artist Ed Ruscha found inspiration from a range of everyday sources, including the iconic Kellogg's "Corny" rooster mascot. These items influenced his work, such as *Rooster* (1987), oil on canvas (**Opposite**).

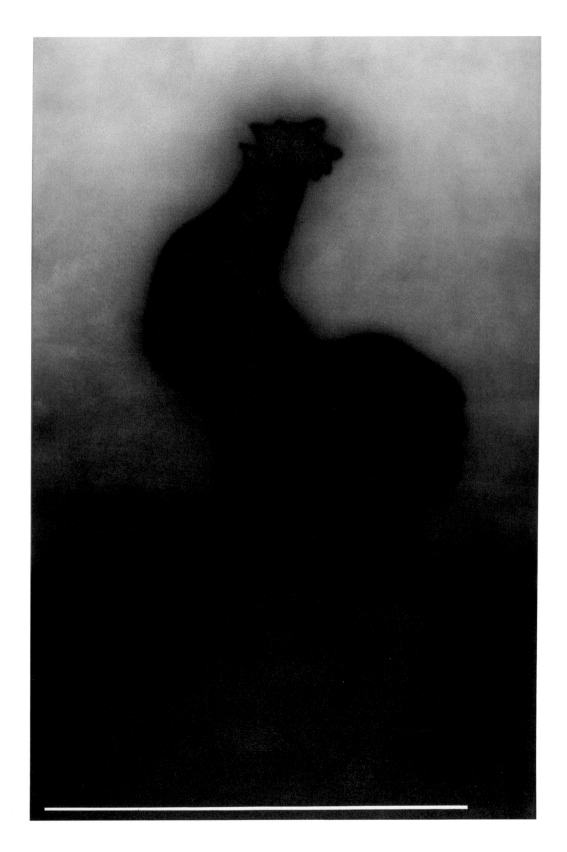

PHOTOGRAPHY AND FILM

Photography of chickens is limited to the years since 1826, when Joseph Nicéphore Niépce produced what is thought to be the oldest surviving photographic image—a view of the barnyard outside his window—giving rise to this now ubiquitous medium. With ever-evolving photographic processes came new ways to capture the charm of backyard bantams, fancy breeds, and their owners. More recently, prize-winning poultry has inspired a host of photographic artworks and books dedicated to an awe-inspiring diversity of breeds, from images produced for Koen Vanmechelen's Cosmopolitan Chicken Project to posed portraits for Stephen Green-Armytage's *Extraordinary Chickens* (2000), Tamara Staples's *The Fairest Fowl* (2001), and Moreno Monti and Matteo Tranchellini's *Chicken* (2018).

Opposite One of the earliest color photographs was taken by Lous Ducos du Hauron, entitled *Still Life with Rooster*, and taken between 1869–79.
Above *Chicken Feathers* (1840) was a stunning early photograph taken by Nevil Story-Maskelyne after WIlliam Fox Talbot, in what is known as a salted paper print.

No. 797. Feeding the Chickens

The albumen effect
(1841–present)

Chickens played a larger role in the history of photography than many people might appreciate. Not only were they subjects of interest (in the farmyard, within portraiture, or as part of the avian world) but they also contributed to the development of the medium itself with the introduction of albumen—more commonly known as egg white—to produce photographic prints.

Invented by Louis Désiré Blanquart-Evrard in 1850—not long after the introduction of the heliograph (the process used by Nicéphore Niépce to produce his barnyard scene), the daguerreotype in 1839, and the calotype in 1841—the albumen print (or albumen silver print) played a vital role in bringing photography to the masses as it procured a sharp, clear image and, due the nature of the process, could be easily reproduced.

Made by coating a thin piece of cotton paper with an emulsion of albumen and salt, the dried and sealed glossy surface was then dipped in a solution of silver nitrate and water, rendering it sensitive to UV light. It was then dried again, this time in the dark, placed in direct contact with a glass negative, exposed to sunlight until the desired level of darkness was acquired on the positive print, and finally bathed in sodium thiosulfate to fix the exposure.

Albumen prints were hugely popular between 1860 and 1890, particularly in the production of soft, sepia-toned *carte de visites* ("visiting cards")—small photographic portraits mounted on card for distributing among friends and family and collecting in albums—and the larger-format cabinet cards that superseded them. Albumen prints eventually fell out of fashion with the advent of gelatin silver prints, carbon printing, and color photography, but there are still numerous charming (if faded and often yellow-stained) examples of them to be found featuring images of chickens and roosters.

Underlining the chicken theme, Victorian photographer Julia Margaret Cameron (1815–79) produced many of her albumen-print, soft-focused, close-up portraits from a studio fashioned out of a glass-roofed chicken coop and a dark room in the coal house. And one of the earliest color photographs not to have been manually colored by hand was of a rooster and a parakeet, attributed to the pioneering physicist and artist Louis Arthur Ducos du Hauron (1837–1920), see page 356.

As photography went mainstream with the introduction of the Kodak Brownie in 1900, Kodachrome film in 1935, and numerous cameras, film types, and then digital formats until the present day, so a fledgling experiment gave way to a worldwide proliferation of images, with chickens running the gamut of works from family snapshots, selfies, photojournalism, still lifes, and portraiture to wildlife, fashion, and street photography.

Opposite Entitled simply *Chicken* (1872), these iconic albumen silver prints were some of the earliest photographs ever taken, seen here as popular visiting cards of the day. **Following spread** *Cote d'Azur, Var, Saint Tropez* (1959) is an iconic work by the famed French photographer Henri Cartier-Bresson, using a technique known as gelatin silver print.

Koen Vanmechelen

(1965–present)

Chickens and their eggs are so woven into the tapestry of modern life that it's easy to take them for granted. Not so for the Belgian multidisciplinary conceptual artist Koen Vanmechelen, who has made it part of his life's work to shine a light on the lack of attention paid to the biodiverse backstory of the domesticated chicken and to prompt much-needed collective awareness.

Vanmechelen is best known for his ongoing Cosmopolitan Chicken Project (CCP), initiated in 1999—a quest to crossbreed domesticated chickens from different countries with the aim of creating a cosmopolitan hybrid that carries the genes of all breeds worldwide. His explorations hold both artistic and community value, having resulted in a proliferation of drawings, paintings, sculptures, and installations, including photography and film, as well as ways to improve chicken health, longevity, and fertility globally via the spin-off Planetary Community Chicken (PCC) project. The former has been exhibited in galleries around the world—most recently as part of the V&A's 2019 show *FOOD: Bigger than the Plate*—while the latter has interacted with scientists as well as local communities and their chickens in purposely established Global Open Farms (real-life crossroads where art, science, and industry meet) in Harare (Zimbabwe), Detroit (US), and Addis Ababa (Ethiopia), each chick born in a local creative space in order to invite public debate.

In 2017, Vanmechelen moved his operations to the 60-acre (24.3-hectare) Labiomista, a collaboration with the Belgian city of Genk established to further explore issues of identity, diversity, and locality. Vanmechelen's studio is located in a building known as "The Battery," and here the chicken rules the roost again, with an installation devoted to the Red Junglefowl, various egg sculptures, and live hens and roosters, among other animals.

From hatching chicks in his bedroom as a child, to his first foray into matchmaking chickens by pairing a Belgian Mechelen Cuckoo with a French Poulet de Bresse (the first CCP), to the many-feathered self-portrait (*Ubuntu*) exhibited at the *Seduzione* exhibition at the Uffizi Gallery (2022), Vanmechelen makes it clear that chickens are a multicultural force to be reckoned with.

Opposite Central to Koen Vanmechelen's Cosmopolitan Chicken Project (CCP) is the idea of crossbreeding domesticated chickens from around the world to create a cosmopolitan hybrid with the genes of all the breeds worldwide. Here, Vanmechelen is seen working with one of his prized collaborators.

Moreno Monti
and Matteo Tranchellini
(1965 & 1968–present)

For anyone averse to the charm of chickens, Italian duo Moreno Monti and Matteo Tranchellini's glamorous curation of fair fowl is sure to turn your eye. Their crowdfunded CHICken project involved taking 200 stunning portraits of 100 different breeds, followed by the release of a self-published collectable photobook entitled *Chicken*, and these not-so-humble birds amaze and allure with their strutting style and awe-inspiring plumage. Inspired by Tranchellini's search for a round, fluffy, feathery-legged pet Cochin for his studio in 2013, the two photographers soon started to attend chicken shows and aviary exhibitions and were blown away by the variety of breeds they saw. Having worked together for years on photographic projects, they persuaded chicken breeders to let them shoot their wares, using only a simple light set-up and encouraging flexibility around "breed standard" poses in order to get the most captivating shots. So popular was the series of bedazzling hens and roosters that the duo followed up with the first and only large-format photobook of poultry, *Chicken: A Declaration of Love* (teNeues, 2020), including details on various breeds and how to look after them.

Opposite The powerfully stunning works of Moreno Monti and Matteo Tranchellini are documented in their crowdfunded CHICken project, composed of 200 chicken portraits. This beautiful pair can be found in the 2020 book entitled *Chicken: A Declaration of Love*.

Stephen Green-Armytage

(1938–present)

As an editorial and commercial photographer working on both sides of the Atlantic, British-born Stephen Green-Armytage honed his style across a range of major advertising campaigns and magazines such as *Sports Illustrated*, *Fortune*, and *LIFE*. It was the latter that, alongside sports shots and portraiture, provided him with a platform for his animal photography, including wild and domestic birds, mammals, and reptiles, so creating books of some of his favorite subjects was a natural progression. The stunning *Extraordinary Chickens* (Abrams, 2000) presented studio portraits of more than fifty flamboyant breeds from around the world, including the Bearded Silkie, the Crested Polish, and the Phoenix. More poultry pin-ups followed in the similarly bijoux revised edition, entitled *Extra Extraordinary Chickens* (Abrams, 2005), showcasing sixty-one breeds (with fourteen not included in the previous book) and five species, as well as sections devoted to combs, feet, crests, and patterned chickens. So extraordinary are these domestic fowl that they now also have an annual parade in the form of the *Extraordinary Chickens* wall calendar (Abrams).

Opposite Stephen Green-Armytage's photographic style was finetuned from his work in advertising and magazines, such as *Sports Illustrated*. What a wonderful path to see that expertise in portraiture wend its way to the world of chickens and other animal photography. His brilliant portraits capture the beauty, majesty, and even humor of these birds, perfectly captured here in this example from the 2020 *Extraordinary Chickens* wall calendar.

Tamara Staples
(1964–present)

For former prop stylist and commercial and fine-art photographer Tamara Staples, the route to chicken photography ran in parallel with time spent with her uncle, Ron, an enthusiastic chicken breeder with his own henhouse, who took Staples to her first ever poultry show in 1988. Around a decade later, a tour of shows around the Midwest eventually resulted in enough photographic fodder for her first book, *The Fairest Fowl: Portraits of Championship Chickens* (Chronicle Books, 2001). A fully revised and expanded edition appeared in 2019, entitled *The Magnificent Chicken: Portraits of the Fairest Fowl* (Chronicle Books, 2019). Dedicated to the fanciers who spend years trying to meet the "Standard of Perfection" for their breed (for example, honing the perfect feet, wingspan, color, and weight) as well as the works of art they create—a diverse array of chickens including the Self Blue Bearded D'Anvers, the White Showgirl Bantam, the Black Langshan, and the White Leghorn Bantam—each portrait aims to bring out the unique personality of the individual bird as well as its collective form.

Opposite Tamara Staples striking chicken portraits successfully captures the unique personality of each individual, while displaying the beauty of the chicken form and breed traits. The beautiful *Black Langshan* seen here is regal in stature but very real in its all-too-realistic uneven foot feathering and dreamy gaze.

Tim Flach

(1958–present)

British photographer Tim Flach's intimate studio portraits of animals—featuring a wide array of species, from pigeons to yellow-eyed tree frogs, opera bats, and labradoodles—go beyond the merely aesthetic to illuminate the relationship between such creatures and humankind and to explore "the role of imagery in fostering an emotional connection." An Honorary Fellow of the Royal Photographic Society, Tim's major bodies of work include *Equus* (2008), *Dog Gods* (2010), *More Than Human* (2012), *Endangered* (2017), and *Birds* (2021)—all of which were also published by Abrams as books, alongside *Evolution* (2013) and a children's book entitled *Who Am I?* (2019). Within this oeuvre exist several images of chickens, including the iridescent green-and-black *Black Japanese Bantam*, the astonishingly grand, fuzzy-hatted *Bantam Silver Laced Frizzle Polish Chicken*, and a pair of long-necked *Showgirl Chickens*, with not a hint of domestic between them. Perhaps the most arresting in show, however, is the controversial *Featherless Chicken*, which portrays a chicken specially bred for its prêt-à-porter, culinary credentials, dancing across the frame like a ballerina, its bright red comb and wattle at least still proudly in place—a strikingly anthropomorphic and anthropocentric reminder of how we shape animals and their meaning.

Opposite Tim Flach's glamorous *Bantam Silver Laced Frizzle Polish Chicken* is truly stunning. The subject so otherworldy, the texture and color so mesmerizing, it may take the viewer a moment to recognize what they are seeing is truly a chicken.

Cartoon chicks

The Looney Tunes cartoon rooster Foghorn Leghorn is probably the most famous of the animated flock, and certainly the most recognized chicken breed—created during the golden age of American animation by Robert McKimson and Warren Foster and featured in some twenty-nine cartoons between 1946 and 1964. Although he's billed as the loudest rooster on the block, there are other cinematic chickens of worth, including his love interest, the hen Miss Prissy. Others include the loyal but feisty Lady Kluck, lady-in-waiting to Maid Marian in the Disney version of *Robin Hood* (1973); Ginger, an alpha hen determined to save her fellow chickens from doom with the help of the rooster Rocky Rhodes in the award-winning Pathé, Aardman Animations, and Dreamworks stop-motion feature *Chicken Run* (2000); Chicken Little from the 2005 movie of the same name, adapted from the old fable "Henny Penny"; and the incongruently billed, easily distracted, and accident prone rooster Hei Hei (meaning "chicken" in Maori), who unknowingly stows away on the adventure-going canoe in Disney's *Moana* (2016). While some of these characters crop up in cartoons or movies that revolve around chickens and thus include whole broods of them, these quirky birds also seem to provide excellent fodder for comedic or cocky turns next to more balanced characters in the form of other animals.

Opposite Chickens have long held a place in our hearts and on our screens, from Looney Toons to *Moana's* Hei Hei. Here, Aardman Animations stars of *Chicken Run* have found themselves dodging a pie adventure.

Fowls on film

Chickens appear in numerous movies, often as supporting characters in the farmyard. One example that introduces them to more art-house effect is the tragicomedy *Stroszek* (1977) by the German director Werner Herzog, which closes with a somewhat dark final scene of a chicken dancing haplessly to a soundtrack of repetitive arcade music— a metaphor for the harsh realities underpinning the pursuit of the impossible quest that is the American Dream. On a lighter note, the documentary *Chicken People* (2016), directed by Nicole Lucas Haimes, follows three attendees of the Ohio National Poultry Show in Columbus as they compete to win prizes for their show-stopping breeds, revealing the subculture that accompanies this often-obsessive activity, and the reasons why people are drawn to it. And an earlier BBC documentary entitled *The Private Life of Chickens* (2010) illustrates an entirely more bucolic scene as presenter and patron of the British Hen Welfare Trust Jimmy Doherty explores the secret lives of chickens on a farm in Devon, endeavoring to uncover how they decide their pecking order, deal with predators, communicate with their chicks, use extensive binocular vision to look in two directions at once, and even go through something known as sex reversal, where a female takes on the traits of a male.

Opposite The Looney Toons character of Foghorn Leghorn is one of the most iconic chickens on screens big and small, as well as the most recognized chicken breed. In addition to appearing in twenty-nine animations, Foghorn Leghorn made a big-screen appearance on film in the 1981 Warner Bros. film *The Looney Looney Looney Bugs Bunny Movie.*

DESIGN, CRAFT, AND STYLE

The domesticated chicken and all it represents is a motif that has stood the test of time, from rooster-themed folk art in the form of paper cuts and figurines (including Oaxacan *alebrijes*, the Portuguese Rooster of Barcelos, and the Swedish Dala Rooster), to centuries of egg offerings at Zoroastrian Nowruz celebrations, cockerel-topped churches (made official by Pope Nicholas I in the ninth century), barnyard textile designs, and many collages and prints from artists like Mark Hearld, Edward Bawden, Eric Ravilious, and Arthur Parkinson. As chicken fancying and rearing becomes popular again, it's a design choice that looks set to stay.

Opposite The traditional Rooster of Barcelos is the national emblem of Portugal (seen here against the backdrop of Porto). **Above** Traditional papercutting is a common folk art form from the Lowicz region of Poland and often includes colorful roosters.

Traditional arts and crafts often involve animals and plants common to
the region. **Left** This papercutting from Poland depicts roosters, wheat,
and flowers, while **Right** This Slavic floral embroidery demonstrates a
traditional folk pattern of similar elements.

Fowls in folk art

If folk art is an expression of the cultural life and identity of a community, chickens are one of its most commonly recurring motifs, with roosters stealing most of the limelight. There are several reasons for this, relating to their inherent traits. The male tends to crow in the morning, signaling hope and associations with warding off the evil spirits of the night. A rooster is also highly territorial and defends its flock vigorously—it has long spurs and will fight to the death, leading to an association with protection, loyalty, valor, and persistence. It enjoys strutting around, displaying its prominent comb and wattles and striking tail feathers, proudly self-confident despite being a largely flightless bird. And with the ability to fertilize up to fourteen eggs at a time, with its sperm lasting up to two weeks inside a hen's reproductive system, the rooster is also celebrated for its virility.

Chickens have been domesticated for thousands of years, so it's no surprise to find that their likeness has populated the folk art of countless civilizations. In Nigeria, we find roosters in the form of what are now known as Benin Bronzes (see page 339), paying homage to the queen mothers of the Edo people. In Mexico, they appear as brightly painted sculptures known as *alebrijes*, carved from the branches of the sacred copal tree (*Bursera glabrifolia*), native to central Mexico. These fantastical creatures were first conceived as by the artist Pedro Linares (1906–92), inspired by a series of feverish hallucinations; taking the form of papier-mâché sculptures, they were much appreciated by Frida Kahlo and Diego Rivera. With

its tradition of wood carving, the craftsmen of Oaxaca later adapted his concept to produce similar sculptures from the local copal.

In Sweden chickens present as the Dala Rooster, similar in style to the painted wooden statue known as the Dala Horse, with both originating from the province of Dalarna, near Mora. Over in Portugal it's the highly stylized Rooster of Barcelos (see page 350), another much-loved national symbol. Meanwhile, Poland has the popular early nineteenth-century folk craft of *wycinanki*—beautiful paper cuts made by hand using a single piece of paper (as practiced in the region of Kurpie) or multicolored, using more than one sheet (the style used in the Lowicz region). Symmetrical in design, these were formerly used by Polish peasants to decorate their homes, and they often featured roosters as symbols of hope. The Chinese, too, use paper cuts to decorate their homes—at Chinese New Year and also for other celebrations—with the rooster considered a symbol of prosperity. And chicken-shaped metal incense burners are a popular tradition in many countries throughout Asia, including India and Indonesia.

Roosters and hens are also prevalent in Native American culture, although some depictions are of the unrelated prairie chicken. Most recently, the work of contemporary Navajo folk artists Edith and Guy John continues this tradition using cottonwood to fashion nesting or pecking figurines.

All in all, a world of chickens can be found in celebration of this most cultural fowl.

Top The brightly painted mythical sculptures of Oaxaca, Mexico, known as *alebrijes* are often made up of a cross between existing animals (such as this large rooster *alebrije*) but with a magical flair. **Bottom** Traditional Chinese papercuttings are common for Chinese New Year and other celebrations, where the rooster is considered a symbol of prosperity.

Odes to eggs

What if the world was born of an egg? This is precisely what many cultures throughout history have believed, including the peoples of India, Egypt, Greece, Phoenicia, China, Polynesia, Persia, West Africa, Finland, Ukraine, and Estonia, the notion of a world, or "cosmic egg," being a logical concept for the "hatching" of life. And with such beliefs came a host of different mythologies and traditions.

Early humans were marking eggshells with symbolic patterns in ancient times, as evidenced by the discovery of 60,000-year-old engraved eggshells at Diepkloof Rock Shelter in South Africa. The Ancient Egyptians hung inscribed, dyed eggs as symbols of renewed life to usher in their harvest season, Shemu (and they were also one of the first civilizations to incubate chicken eggs). In Mesopotamia, red-dyed eggs symbolized the blood of Christ—one possible origin of the custom of Easter eggs, and painted eggs have long been used to celebrate the day of the Persian New Year, Nowruz, on the spring equinox. Boiling eggs with onion skins to produce patterns on their shells was practiced in pagan times in both Scandinavia and Northern England, and the beautifully decorated batik-dyed, appliquéd, scratched, waxed, and carved eggs of Russia, Ukraine, Lithuania, and Poland likewise date back to Slavic paganism. And in Mexico and Mesoamerica, egg-cleansing ceremonies evolved in which a whole egg was rolled over the body to cleanse it of evil spirits.

Perhaps the most famous decorated eggs of all were created by the Russian House of Fabergé (see page 319). Established by Gustav Fabergé in St. Petersburg in 1842, the eggs were produced during the stewardship of his son Peter Carl Fabergé, who transformed the company into artist jewelers famous for their exquisite goldwork, colorful gemstones, and for reviving the lost art of enameling. In 1885, Czar Alexander III commissioned the company to make an Easter egg for his wife, Maria Feodorovna. Initially intended to contain a diamond ring, what is now known as the "Hen Egg" was inspired by an eighteenth-century objet d'art. It consisted of an opaque white enameled outer "shell" that twisted open to reveal the first surprise—a matte yellow gold yolk. Then, inside the yoke was a gold enameled hen that once contained a replica of the Imperial Crown and a precious ruby egg.

Although the Hen Egg is the simplest of all the Imperial Eggs that followed—fifty in total, all unique and increasingly elaborate in style—it is perhaps the most beautiful for its pared-back homage to the powerful life force of the egg.

The Karelian Birch Egg and the Blue Constellation Egg, both made in 1917, were never delivered to Czar Nicholas II (intended for his mother Maria and wife Alexandra respectively) due to the outbreak of the 1917 February Revolution and the Czar's forced abdication. The entire family was killed by the Bolsheviks in 1918, leaving the richly decorated eggs as a legacy of a bygone, pre-Soviet era.

Opposite (clockwise from top left) Eggs and their shells have held a special place in the mythologies, traditions, and art forms of cultures throughout history and around the world. In Eastern Europe, decorated eggs are seen as a sign of rebirth around the time of Easter; here, the Romanian artist is using beeswax and dyes. Eggs are dyed in a wide range of ways around the globe. Flower pressing, batik, carving, waxing, or natural or artificial dyeing are all ways to demonstrate symbolic mark-making on eggshells. **Following spread** Egg dyes and paints can be made using natural plant-based dyes, seen here with red onion peels for coloring.

Weathercocks

Weather vanes—placed on top of buildings to tell the direction of the wind—are an ancient tradition thought to date back at least as far as Ancient Greece and the Vikings. Many of them are fashioned in the form of cockerels, including the earliest-known weather vane, the plump golden effigy known as the Gallo di Ramperto (Rooster of Ramperto), thought to have been made between 820 and 830 in Brescia, Italy, where it roosted atop a church tower for over a thousand years before being moved for safekeeping to the nearby Museum of Santa Giulia. Although birds had been used as weather vanes before—on the roofs of Ancient Chinese palaces, for example— weathercocks became intrinsically associated with churches after Pope Leo IV (790–855) placed a cockerel on top of Old St. Peter's Basilica. Inspired by Pope Gregory I (*c.* 540–604), who had declared the rooster the most suitable emblem of Christianity— signifying Saint Peter who, according to the Bible, would deny Jesus three times "before the rooster crows"—cocks were eventually ordered to be placed on every church steeple by Pope Nicholas I (*c.* 800–867) as a reminder of the importance of faith. They can now be found around the world, wherever the Church has left its mark.

Weather vanes have long displayed chickens prominently from the rooftops of the world, including the earliest known, Gallo di Ramperto (Rooster of Ramperto) from 820-830 in Italy. **Right** This simple weather vane is from a barn in New Jersey, USA. **Opposite** This antique wrought-iron weather vane is a beautiful example of this long-standing tradition. **Following spread** This strikingly abstract weather vane is found in Tallinn, Estonia on a frosty winter morning.

Arthur Parkinson
(1993–present)

Chicken fancying is enjoying a new heyday, thanks to prominent hen lovers like small-space gardening sensation and writer Arthur Parkinson, who celebrates his favorite birds (a passion since childhood)—including Marans, Cream Legbars, Chamois Polish Bantams, Pekin Bantams, and Buff Cochins—through an ensemble of dedications, painted portraits, and photographs. Having studied horticulture at the Royal Botanic Gardens of Kew, Parkinson worked with gardener Sarah Raven, honing his passion for cut flowers. Arthur owes his passion for chickens to the late Deborah Devonshire's collection of poultry that she kept on public view while she was the tenth Duchess of Devonshire at Chatsworth in Derbyshire, UK. Arthur and the Duchess developed an annual correspondence on the subject of their mutual passion for chickens, and he credits her for giving style and attention to a number of rare breeds. During the UK COVID-19 lockdown of 2020, Arthur cared for his nan, Minnie Brown. They lived together with a number of bantams and appeared on BBC Radio 4's *You and Yours* program, discussing the comfort of chicken-keeping.

Arthur's first book, *The Pottery Gardener: Flowers and Hens at the Emma Bridgewater Factory* (The History Press, 2018), celebrated the colorful haven he created there, a paradise for visiting pottery lovers and the chickens that roamed around the yard. His next book, *The Flower Yard: Growing Flamboyant Flowers in Containers* (Kyle Books, 2021), included an array of hens peeking between the tulips, dahlias, peonies, and compost piles, including a Belgian Barbu d'Uccle Millefleur bantam hen and a frizzle-plumed Buff Pekin. A fully dedicated chicken book is up next, with the working title *Chicken Boy: A Life of Hens* (2023)—part memoir, part guide to keeping hens at home, with charmingly illustrated starring roles by Buff Orpingtons and Belgian bantams as well as rescued battery hens.

Opposite (clockwise from left) These charming and characterful original chicken illustrations by Arthur Parkinson include the Appenzeller Spitzhauben, Buff Orpington, Cream Legbar, and Speckled Sussex.
Above Captured by photographer Jonathan Buckley, Arthur Parkinson can be seen here doing what he loves best—surrounded by chickens and gardens.

Mark Hearld
(1974–present)

The fauna and flora of the British countryside are central to Mark Hearld's work, from distinctive mixed-media collages to lithographs, linocuts, paintings, ceramics, and wallpaper and textile designs. Among the foxes, owls, blackbirds, ducks, and squirrels there are also numerous portrayals of the domesticated chicken, from flamboyant roosters in *Rooster and Railway Carriage* (2009) and *Remembered Farm* (2020) to a red-and-blue linocut of a *Cockerel* (2020) and a fold-out collage of the whole family in the greeting card "Chickens." On being asked to help curate and rehang the British Folk Art Collection at Compton Verney, Warwickshire, in 2021, Hearld also produced a new body of work, including the *Compton Verney Collage*, with hens, chicks, and roosters situated alongside other farmyard animals and plants to represent the grounds of the Compton Verney manor and surrounding countryside, plus a metal cutout of a cockerel silhouetted against a window. His work is so sensitively observed and produced—his technique and style having been honed through an MA in natural history illustration at London's Royal Collage and Art, and influences including similarly illustrative and chicken-loving artists and printmakers Edward Bawden (1903–89) and Eric Ravilious (1903–42)—that simply looking at it feels like you're right there in the barnyard.

Above Eric Ravilious' wood engraving of *Tirzah on a Cockerel* (1931) was used as an emblem of the Golden Cockerel Press as its front cover. **Right** Seen at an exhibition of his work in Edinburgh, Scotland, Mark Hearld here stands in front of *Remembered Farm* (2020).

ADDITIONAL
RESOURCES

Dedicated to Lily

~

To all the chickens around the world who have nobly provided love, life, sustenance, and unending joy and laughter to chicken keepers and enthusiasts everywhere.

Opposite The beloved children's book series of "Pettson and Findus," from Swedish author Sven Nordqvist, featuring old Farmer Pettson and his antics with his cat Findus, a ramshackle farm, and a cast of exuberant chickens.

ACKNOWLEDGMENTS

The contributors wish to thank...

Jessica Ford would like to thank her husband, Ryan, for his unwavering support and expertise, and Randy Schultz for introducing her to the wonderful world of chicken writing. A special thank you to mom, Patti, and Grammie Bear for giving her a lifelong love of birds. Many thanks as well to her two sweet boys and her whole happy flock of backyard chickens. Finally, Jessica extends her deepest thanks to Caitlin Doyle for the opportunity to contribute to this book.

Sonya Patel Ellis wishes to thank all the artists who incorporated chickens into their work, a far greater and more fascinating oeuvre than one might at first imagine. To Sylvester and Iggy who came on the chicken journey with me and ate lots of eggs along the way. But mostly, dedicated to beautiful, brave Lily, who loved her chickens, inspired all those around her, and who I will think of every day.

Rachel Federman would like to thank the ever high-flying Caitlin Doyle for including me in this charming project. Should I admit I never knew chickens flew? Thanks to the Key West Wildlife Center for their round-the-clock rescues and to my dad for inspiring me to become a vegetarian at age six.

The publisher wishes to thank Michael Sand at Abrams for his dedicated and enthusiastic collaboration on such a beautiful series and Soyolmaa Lkhagvadorj at Abrams for her warm and conscientious attention to detail; Ellie Ridsdale Colussi for her stunning design and hard work once again; Milena Harrison-Gray for going above and beyond in her picture research; Lynn Hatzius for her gorgeous illustrations and step-by-steps; Helena Caldon and Rachel Malig for their excellent editorial work; Geraldine Beare for the thorough index; Chris Wright for ensuring the book actually made it to print; and Myles Archibald and Hazel Eriksson for their support and teamwork. Immense gratitude goes to Jessica Ford, Sonya Patel Ellis, and Rachel Federman for their expertise and eloquence. You three have been an invaluable asset and pleasure. Thank you to Ryan Ford for the wonderful projects and lovely photographs. Team Chicken has been a joy to work with, and we've all so enjoyed the excuse to pore over countless stunning chicken images in the making of this book. And finally, a big thank you to each and every contributor to this incredible collaborative effort.

FURTHER READING AND RESOURCES

ARTICLES

Adler, Jerry and Andrew Lawler. "How the Chicken Conquered the World." *Smithsonian Magazine*, June 2012. smithsonianmag.com /history/how-the-chicken-conquered-the -world-87583657/

Hata, A., Nunome, M., Suwanasopee, T. et al. "Origin and Evolutionary History of Domestic Chickens Inferred from a Large Population Study of Thai Red Junglefowl and Indigenous Chickens." Sci Rep 11, 2035 (2021). https://doi .org/10.1038/s41598-021-81589-7

The Humane Society of the United States. "Adopting and caring for backyard chickens." humanesociety.org/resources /adopting-and-caring-backyard-chickens

Lovette, Irby. "How Did the Chicken Cross the Sea?" All About Birds, January 2008. The Cornell Lab. allaboutbirds.org/news /how-did-the-chicken-cross-the-sea/

BOOKS
General

Aschwanden, Christie. *Beautiful Chickens: Portraits of Champion Breeds* (Ivy Press, 2020).

Barber, Dr. Joseph, ed. *The Chicken: A Natural History* (Ivy Press, 2012).

Damerow, Gail. *The Chicken Encyclopedia: An Illustrated Reference* (Storey Publishing, 2012).

Polzin, Jackie. *Brood: A Novel* (Doubleday, 2021).

Sandri, Barbara and Francesco Giubbilini. *Chickenology: The Ultimate Encyclopedia* (Princeton Architectural Press, 2021).

Children/poetry

Asim, Jabari. *Preaching to the Chickens: The Story of Young John Lewis* (Nancy Paulsen Books, 2016).

Hwang, Sun-mi. *The Hen Who Dreamed She Could Fly: A Novel* (Penguin, 2013).

Miller, Sara Swan. *Chickens* (Children's Press, 2000).

Milway, Katie Smith. *One Hen: How One Small Loan Made a Big Difference* (Kids Can Press, 2020).

Polacco, Patricia. *Chicken Sunday* (Philomel Books, 1992).

Prelutsky, Jack. "Last Night I Dreamed of Chickens," 1940.

Reed, Avery. *Backyard Chickens* (Penguin Younger Readers, 2015).

Sklansky, Amy E. *Where Do Chicks Come From?* (HarperCollins, 2005).

Steele, Lisa. *Let's Hatch Chicks!: Explore the Wonderful World of Chickens and Eggs* (Young Voyageur, 2018).

Wahl, Phoebe. *Sonya's Chickens* (Penguin Random House, 2015).

Wiberg, Karen. *Chicken Haiku* (Clear Sight Books, 2018).

Chicken Keeping
Caughey, Melissa. *How to Speak Chicken: Why Your Chickens Do What They do & Say what They Say* (Storey Publishing, 2017).

Damerow, Gail. *Storey's Guide to Raising Chickens* (Storey Publishing, 2017).

Kuo, Anne. *The Beginner's Guide to Raising Chickens: How to Raise a Happy Backyard Flock* (Rockridge Press, 2019.)

Rossellini, Isabella. *My Chickens and I* (Abrams Image, 2018).

Steele, Lisa. *Gardening with Chickens: Plans and Plants for You and Your Hens* (Voyageur Press, 2016).

Warren, Gina G. *Hatched: Dispatches from the Backyard Chicken Movement* (University of Washington Press, 2021).

WEBSITES/MAGAZINES/JOURNALS
BackYard Chickens
backyardchickens.com

Backyard Poultry
backyardpoultry.iamcountryside.com

The Cape Coop Farm
thecapecoop.com

Chickens magazine

Chicken Whisperer Magazine
chickenwhisperermagazine.com

Community Chickens
communitychickens.com

The Happy Chicken Coop
thehappychickencoop.com

Hobby Farms: Chickens
hobbyfarms.com/animals/poultry/

Homesteading: Raising Backyard Chickens
homesteading.com/raising-backyard
-chickens

iChicken
ichicken.ca

Journal of Applied Poultry Research
journals.elsevier.com/journal-of-applied
-poultry-research

My Pet Chicken
mypetchicken.com

Poultry Science
sciencedirect.com/journal/poultry-science

Small and Backyard Poultry
poultry.extension.org

APPS

Cluck–ulator
play.google.com/store/apps/details?id=com
.chickenwaterer.p5618jj&hl=en_US&gl=US

FlockPlenty—Chicken Egg Tracker
apps.apple.com/us/app/flockplenty
-chicken-egg-tracker/id1017524534

PODCASTS

Backyard Poultry with the Chicken
Whisperer
blogtalkradio.com/backyardpoultry

Pastured Poultry Talk
pasturedpoultrytalk.com

The Urban Chicken Podcast
urbanchickenpodcast.com

ORGANIZATIONS TO SUPPORT

Chicken Run Rescue
chickenrunrescue.org

Compassion in World Farming, Inc.:
"The Better Chicken Initiative"
www.ciwf.com

HenPower
equalarts.org.uk/our-work/henpower

Penelope's Place—The Sanctuary
penelopesplacethesanctuary.com

Safe Haven Farm Sanctuary
safehavenfarmsanctuary.org/our-animals
/chickens/

Triangle Chicken Advocates
trianglechickenadvocates.org

INDEX

Page numbers in **bold** refer to illustrations

H

hackle feathers 22
hackles 30
Haimes, Nicole Lucas 375
Hamburg **47**, 164, **165**
happiness 222
Harry, Prince, Duke of Sussex 42
hatching 196, 198
Havelock, Henry 340
Hawkins, Benjamin Waterhouse 331
hawks 218
hawthorn 263
hazel 263
health 208–15, 222
Hearld, Mark 376, 392
heat, protection from 226, 229
heat sources 243, 246
hedges 260, 263, 266
Helianthus annuus (black oil sunflower) 270
Helianthus giganteus (giant sunflower) 270
Hemingway, Ernest 45
hen bullying 205–6
Hen Fever (or The Fancy) 26, 38, 63, 323
hens 22, 239
herbs 234, 269
Heritage Chicken breeds 323
Herzog, Werner 375
Himalayacalamus hookerianius (blue bamboo) 263
histoplasmia 210
Hokusai, *Great Wave* 326
Hollar, Wenceslaus 332
Hondecoeter, Gijsbert d' 336
Hondecoeter, Gillis d' 336
Hondecoeter, Melchior d' 336
hornbeam (*Carpinus betulus*) 263
housing 239
Hunt, Edgar 336
Hunt, Walter 336
Hurley, Elizabeth 41
hybrid 22

I

Indian grass (*Sprghastrum nutans*) 269

injuries and ailments, common 213–15
ISA Brown **47**, **132**, 133

J

Jackson, Chris 42
Jakuchu 326
Japanese art 326
Japanese Bantam **47**, **160**, 161, 163
Japanese rose (*Rosa rugosa*) 263
Jefferson, Thomas 26
Jersey Giant **46**, **72**, 73
John, Edith 380
John, Guy 380
Ju Chao 326
junglefowl 331
   see also named junglefowl eg. Red Junglefowl

K

Kahlo, Frida 380
kale (*Brassica oleracea*) 270
Kellogg's 354
Kentucky bluegrass (*Poa pratensis*) 269
Key West, Florida 45
Key West Gypsy Chickens **44**, 45
Key West Wildlife Center 45
kitchen scraps 235
Klimt, Gustav 334, 346
Kodak 359

L

Langshan **46**, **92**, 93
large clumping bamboo (*Bambus* spp.) 263
lash egg (salpingitis) 213
Lavandula angustifolia (English lavender) 269
Lavandula stoechas (French lavender) 269
lavender 269
layer 22
layer feed 22
laying 204
leafy greens 270
Lederer, August 346
Lee, Henry 323
leg, splay (or spraddle) 215

Legbar **46**, **110**, 111, 391
Leghorn **47**, 59, **118**, 119
lemon balm (*Melissa officinalis*) 269
Leo IV, Pope 386
Lewis, Celia, *The Illustrated Guide to Chickens: How to Choose Them, How to Keep Them* 42
Lewis, Earl 41
Lewis, John 38, 40
lice 213
life cycle 36–7
lights 244, 246
Linares, Pedro 380
listeria 210
litter, outdoor 246
little bluestem (*Schizachrium scoparium*) 269
location 239, 253–4
   cleanliness 254, 256
   living in harmony 256–7
   safety 254
Lodge, George Edward 331
Lolium multiflorum (annual ryegrass) 269
Looney Tunes cartoons 372
Ludlow, Joseph Williamson 323

M

McCloskey, Robert, *Make Way for Ducklings* 38
McKimson, Robert 372
Maes, Eugène Rémy 336
main tail 31
Malus angustifolia (southern crab apple) 266
Malus domestica (edible apple) 266
Malus sylvestris (European crab apple) 266
Marans 42, **47**, **112**, **113**, 391
Marie Antoinette 38
marigold 234, 269
marjoram (*Origanum majorana*) 269
Markle, Meghan, Duchess of Sussex 42
mating 195
meat birds *see* broiler
Mechelen Cuckoo 362
Melissa officinalis (lemon balm) 269
melon (*Cucumis* spp.) 270

PICTURE CREDITS

All reasonable efforts have been made by the authors and publishers to trace the copyright owners of the material quoted in this book and of any images reproduced in this book. In the event that the authors or publishers are notified of any mistakes or omissions by copyright owners after publication, the authors and publishers will endeavor to rectify the position accordingly for any subsequent printing.

For thumbnail photographs on pages 46-47, please refer to the main chicken identification pages.

Key: t: top, b: below, m: middle, l: left, r: right

Adobe Stock: 136b, 139t, 140t, 141;

Alamy: 40tr, 40br, 49, 50, 65, 66, 67t, 81, 88t, 89, 90-91, 97, 98t, 99, 106, 112t, 113, 122-123, 126, 128, 138, 143, 147, 148, 159, 192, 194b, 205b, 208t, 212, 214t, 219, 223t, 226b, 230, 235, 237tl, 242t, 258-259, 288, 314-315, 319tr, 320b, 321, 326t, 327, 328-329 / The Natural History Museum 330 / BTEU/RKMLGE 333t, 335, 337, 347, 348 / Eric Murphy 350-351 / Electric Egg Photo 352-353 / Courtesy Everett Collection Inc 373 / Courtesy Everett Collection Inc 374, 376, 377, 378, 379, 382b, 384-385, 386b / Sally Anderson 393, 394-395, 398, 400, 404, 412;

© **Alyssa Pharr via Instagram @pharrcydefarm:** 134, 135t;

© **Arthur Parkinson:** 390 (all);

Bridgeman Images: 26, 39, 42, 325 / Photo © Christie's Images / © Succession Picasso / DACS, London 2022 342, 344, 392b;

Creative Commons: 356b;

© **Ed Ruscha:** 355;

The Elisha Whittelsey Collection: The Elisha Whittelsey Fund, 1963. Courtesy of the Metropolitan Museum of Art 332t;

Getty Images: 12t, 14t, 28-29, 33t, 34-35, 40bl, 40tl, 41bl, 80t, 85m, 102, 110, 118, 129t, 130t, 131, 144, 145t, 146t, 151, 200-201, 206t, 220tl, 225t, 237b, 245tr, 319b, 322, / Werner Forman 338, 381b, 382tl, 388-389;

iStock: 93t;

J. Paul Getty Museum: Courtesy of The J. Paul Getty Museum, Los Angeles 358t, 358b;

© **Jonathan Buckley:** 391t;

© **Kara Hagedorn:** 43t, 43b;

Kellogg Company: 354b;

© **Koen Vanmechelen:** 363;

The Livestock Conservancy: Jeannette Beranger 184t;

© **Lynn Hatzius 2020:** Cover illustrations, 36, 275, 278 (all), 279 (all), 280 (all), 281b, 282 (all), 283 (all), 284 (all), 289, 290 (all), 291b, 295, 296 (all), 297b, 301, 302b, 303 (all), 308b, 309t, 310b;

Magnum Photos: © Henri Cartier-Bresson © Fondation Henri Cartier-Bresson 360-361;

Mary Evans Picture Library: 39, 324;

The Met Museum: 333, The Rubel Collection, Purchase, Lila Acheson Wallace and Anonymous Gifts, 1997. Courtesy of the Metropolitan Museum of Art 357m;

© **Moreno Monti & Matteo Tranchellini:** 364;

Murray McMurray Hatchery: Courtesy of Murray McMurray Hatchery 115

New York Public Libraries Digital Collections, Courtesy of: 331;

Opal: / Sven Nordqvist 397;

© **Ryan Ford:** 227, 228, 252, 274, 277, 285, 294, 298, 299, 300, 304b, 305, 306, 415;

Shutterstock: 4, 6, 7m, 8b, 9, 10-11, 15b, 16, 17, 18l, 21, 23l, 23r, 24-25, 27t, 27b, 30-31, 33b, 44, 45t, 45b, 51, 52, 53t, 54t, 55, 56-57, 58, 59t, 60t, 61t, 61b, 62, 63t, 64t, 68t, 69, 70, 71t, 72, 73t, 74t, 75, 76-77, 78, 79t, 82, 83t, 84t, 86, 87t, 92, 94-95, 96t, 100b, 101m, 103t, 104t, 105, 107t, 108t, 109, 111t, 116b, 117, 119t, 120t, 121, 124b, 125, 127t, 132, 133t, 137m, 142b, 149t, 150t, 152, 153t, 154-155, 156t, 157, 158t, 160, 161t, 162, 163t, 164t, 165m, 166t, 167, 168-169, 170, 171t, 172t, 173, 174, 175t, 176t, 177, 178-179, 180t, 181, 182t, 183, 185, 186t, 187, 188-189, 190-191, 194t, 197, 198t, 198b, 199, 202t, 202b, 207, 209b, 216-217, 220tr, 220bl, 220br, 225b, 231, 233b, 240-241, 242t, 242b, 245tl, 245b, 247, 248-249, 250-251, 252t, 255tl, 255bl, 255br, 257t, 261, 262tl, 262tr, 262bl, 262br, 263-264, 267tl, 267tr, 267b, 268tl, 268tr, 268b, 271tl, 271tr, 271b, 272-273, 286-287, 292-293, 311, 312-313, 316, 317m, 341, 381t, 382tl, 387, 416;

© **Stephen Green-Armytage:** 367;

Sunshine Mesa Farm: 114;

© **Tamara Staples:** 368;

Tim Flach Photography: © Tim Flach, from Birds, 2021 371;

Vlisco: Design © Vlisco B.V. 319tl;

Wiki Commons: 334b.

CONTRIBUTOR BIOGRAPHIES

Jessica Ford is a writer, mother, life-long keeper of chickens, former competitor at American Poultry Association shows, and the chicken and homestead contributor to *Home, Garden and Homestead*—an online "Guide to Modern Living" for creating an independent, healthy, and sustainable homestead lifestyle. Jessica is also a self-professed "chicken nerd at heart." She currently works as an author and digital marketing manager in Colorado Springs, where she lives with her husband, two children, and a small flock of chickens.

Sonya Patel Ellis is a writer, editor, and artist exploring the botanical world and the interconnectedness of nature and culture. Her books include *The Botanical Bible* and *The Backyard Birdwatcher's Bible* (Abrams 2018 and 2020)*, The Heritage Herbal* (British Library Publishing, 2020), and *Nature Tales: Encounters with Britain's Wildlife* (Elliott & Thompson, 2011). She is coauthor and botanical consultant on *The Backyard Chicken Keeper's Bible.*

Rachel Federman is a writer, environmental activist, and nonprofit consultant who has written more than twenty nonfiction books for adults and children, including *The Mindful Gardener* in conjunction with the New York Botanical Gardens (Clarkson Potter, 2017) and *Test Your Dog's IQ* (HarperCollins, 2016). She lives in New York City.